CONSTRUCTIVE DIALOGUE MODELLING

SPEECH INTERACTION AND RATIONAL AGENTS

Wiley Series in Agent Technology

Series Editor: Michael Wooldridge, *University of Liverpool, UK*

The 'Wiley Series in Agent Technology' is a series of comprehensive practical guides and cutting-edge research titles on new developments in agent technologies. The series focuses on all aspects of developing agent-based applications, drawing from the Internet, Telecommunications, and Artificial Intelligence communities with a strong applications/technologies focus.

The books will provide timely, accurate and reliable information about the state of the art to researchers and developers in the Telecommunications and Computing sectors.

Titles in the series:

Padgham/Winikoff: Developing Intelligent Agent Systems 0-470-86120-7 (June 2004)

Bellifemine/Caire/Greenwood: Developing Multi-Agent Systems with JADE 0-470-05747-5 (February 2007)

Bordini/Hübner/Wooldrige: Programming Multi-Agent Systems in AgentSpeak using Jason 0-470-02900-5 (October 2007)

Nishida: Conversational Informatics: An Engineering Approach 0-470-02699-5 (November 2007)

CONSTRUCTIVE DIALOGUE MODELLING

SPEECH INTERACTION AND RATIONAL AGENTS

Kristiina Jokinen
University of Helsinki, Finland

WILEY

A John Wiley and Sons, Ltd., Publication

Registered office
John Wiley & Sons Ltd, The Atrium, Southern Gate, Chichester, West Sussex, PO19 8SQ, United Kingdom

For details of our global editorial offices, for customer services and for information about how to apply for permission to reuse the copyright material in this book please see our website at www.wiley.com.

Library of Congress Cataloging-in-Publication Data

Jokinen, Kristiina.
   Constructive dialogue modelling: speech interaction and rational agents/Kristiina Jokinen.
      p. cm.
   Includes bibliographical references and index.
   ISBN 978-0-470-06026-1 (cloth)
   1. Human-computer interaction. 2. Automatic speech recognition. 3. Intelligent agents
   (Computer software) 4. Dialogue–Computer simulation. I. Title.
   QA76.9.H85J65 2009
   004.01'9–dc22

                                                                        2008046994

A catalogue record for this book is available from the British Library.

ISBN 978-0-470-06026-1(H/B)

Typeset in 11/13 Times by Laserwords Private Limited, Chennai, India
Printed and bound in Great Britain by CPI Antony Rowe, Chippenham, Wiltshire

Contents

Foreword

Although we might like to believe otherwise, the truth is that computers are strange and difficult beasts. The most beautiful user interface in the world can at times be baffling, leaving us bewildered and confused. The most carefully engineered program can crash, leaving us frustrated and angry. And of course, all these problems stem from the fact that computers are not like me and you: although we dress them up to appear friendly, colourful, and helpful, they ultimately understand the world only as strings of 1s and 0s. There is a fundamental barrier to communication between human and machine. To fully exploit the potential of computers, we need machines that can relate to us in our terms; that can model, understand, and make use of the methods of communication that we use as humans. The present text addresses exactly these issues. It presents a thorough overview of the area of interaction, focusing particularly on the idea of interaction as a rational process, and the notion of rational dialogue. As well as considering the underlying principles, the book makes a solid contribution to the engineering of dialogue and interaction systems. It represents a valuable step in the path to building systems that can overcome the fundamental barrier between human and machine.

Michael Wooldridge

Preface

State-of-the-art speech and language technology has reached a level that allows us to build interactive applications which the users can have short conversations with in order to search for information. We are already dealing with electronic banking facilities, information providing systems, restaurant guides, timetable services, assisting translation systems, emails, web browsers, etc., which can understand the users' speech and which can also reply using speech. Speech-based interfaces have also brought forward novel applications for situations where interactions using keyboard or mouse clicks are cumbersome or are not possible at all: car navigation, telephone services, home appliances, etc. Moreover, they also provide solutions for users for whom the ordinary mouse and keyboard would not be a possible means of interaction with computer services, thus contributing to the requirement of universal access to digital services and databases. However, the challenge that present-day speech and language research faces in an ever-expanding information society is not so much in producing tools and systems that would enable interaction with automatic services in the first place, but rather, to design and build systems that would allow interaction to take place in a natural way.

This book is about natural interaction in dialogue management. It introduces the Constructive Dialogue Modelling (CDM) approach to interaction modelling, based on a view of dialogue as a shared activity. CDM focuses especially on the aspects of rational and cooperative communication that allow humans to transmit, exchange, mediate, argue and ask for information in an efficient and flexible manner. The CDM approach provides an account of various issues concerning cooperative communication, in which the participants together construct dialogues by exchanging new information on a particular topic in a given context.

The book is not about communication and debate, rhetoric or persuasion, and it does not teach successful conversational styles or strategies. Rather, it focuses on the conversational features and preconceptions that would make the interaction between humans and computers more natural and intuitive. From the interaction design point of view, the goal can be expressed as increasing the usability of speech-based interactive systems: making interfaces more usable and intuitive to use, giving extra value to the system by enabling it to converse naturally with the user.

In order to achieve this goal, we need to equip systems with a component that takes care of interaction management and uses a special model to extract information that goes beyond the compositional meaning of the observed utterances and words. In other words, we need to abstract the intended meaning from the observed utterances and model the speaker's intentions and goals and roles in the activities the speaker is engaged in. Deliberations about the next dialogue action take place in this kind of intention space, and the observed utterances function as a kind of a trigger for the speakers to update the intention space appropriately. This means that usability and user-friendliness in speech-based interactive interfaces are related to the system's communicative capabilities. The users' perception of the system depends on the system's communicative capabilities, which affect the user's view of how satisfactorily the system assists the user in the task that it is meant to accomplish, and how much the system's operation can be trusted. Natural communication appears to support user satisfaction even if the interaction would contain such obviously undesirable features like long waiting times and minor errors: it provides a means to resolve misunderstandings by negotiation and talking. Thus it becomes possible to build systems that function in a more satisfactory manner from the user's point of view: interactions are perceived by the user as useful for the task at hand. It is the system's intelligent interaction that provides the basis for good interfaces and services.

The book provides a concise overview of the different interactional models as well as concepts that enable us to build practical interactive systems, and to test hypotheses for friendly and flexible human-computer interaction, with special attention paid to multimodality. It is intended for communication researchers and computer scientists aiming to design complex interactive systems and experimenting with various types of complex systems using natural language. Taking the view that computers are not only tools but agents that users need to interact with, the possibilities for making interaction more flexible and natural by taking some human language capabilities into account are examined. In particular, the focus is on the aspects of interaction that contribute to smoothness of communication, and on adaptation of the system to the user's level of expertise.

It has often been argued that theoretical approaches to dialogue management produce descriptive models which do not necessarily address concrete problems in interaction technology, while practical approaches provide heuristic and ad-hoc solutions which are not easy to extend to other domains or applications. The CDM model attempts to bring the two views together by insisting on a theoretical rather than add-hoc basis for modelling, and on proof-of-concept experimentation and open evaluation with different systems. The theoretical part of the CDM model is drawn from the existing approaches to dialogues, especially from dialogue planning and rational agency, and from empirical research on human-human communicative behaviour. Practical aspects are related to various research projects where the CDM model has been applied and partially implemented.

Content of the Book

The content of the book can be divided into two parts. The first two chapters form the introductory part where the background and the starting point are introduced. The chapters provide an overview of dialogue management and the current state of the art in general. The second part of the book provides a more detailed account of the CDM approach and various topics related to its development, including discussion of some future views of cooperative, rational dialogue management. In detail, the book is organised as follows.

After the introduction to human-human and human-computer communication in Chapter 1, Chapter 2 introduces different dialogue management models and dialogue systems as well as system architectures and representations. It also highlights the problematic areas of current state-of-the-art interactive systems so as to lay a foundation for the following chapters. The work done in different projects on adaptive speech-based human-computer interaction is also reviewed, providing an overview of the various interactive systems. Chapter 3 then proceeds by presenting the Constructive Dialogue Model (CDM) and the basic concepts that deal with cooperation, rationality and adaptation. Chapter 4 discusses a few examples of the implementation of the basic concepts in dialogue systems. The book then continues with two specific topics: Chapter 5 discusses the management of dialogue and domain knowledge which is necessary for a dialogue system in order to exhibit intelligent interaction, and Chapter 6 focuses on the learning and adaptation necessary for the system to operate and manage interactions in dynamic contexts. Finally, Chapter 7 concludes with future challenges, and contains a roadmap of the interactive systems that we might expect to see in ten years' time.

Introduction

In this chapter, the main objectives of the book are introduced. The new metaphor for human-computer interaction, i.e. the computer as an agent as opposed to a tool, is presented and discussed. The emphasis is on the view of the dialogue as "dialogue in natural language", as opposed to "dialogue by icon clicking". The challenges and opportunities that this kind of interaction bring to speech-based application design are discussed, in particular the need for the modelling of rationality, agenthood, cooperation and natural language interaction. Speech adds much more than just a sound to interactive applications, and presupposes that the different aspects are addressed properly. The main claim of the book is also put forward in the chapter: that natural language communication is the genuine human factor necessary for building flexible and intelligent interactive systems Combined with the agent metaphor, this claim will be substantiated in the Constructive Dialogue Model.

Dialogue Models

This chapter provides the background for the development of CDM. It surveys various dialogue models and their implementation as dialogue management engines, and also gives a short history of (natural language) interactive systems and dialogue management development as opposed to point-and-click interface design. The systems are highlighted as examples of the ideas which have functioned as sources of inspiration for CDM rather than providing a comprehensive overview of the field. Dialogue modelling and practical system building are contrasted, so as to go back to the goals of Chapter 1 in order to argue that we need a deeper understanding of rationality and natural language conversation modelling so we can develop intelligent applications.

Constructive Dialogue Model

In this chapter, the framework of the Constructive Dialogue Model is presented. The basic principles related to the activity-based analysis of communication, rational agency, and to the concept of Ideal Cooperation are discussed. The construction of shared context and mutual knowledge are also studied, with reference made to the previous work on grounding. The possibilities for enabling conversations with computer agents are investigated, and special attention is paid to the socio-cultural context in which the dialogues take place. The focus is on dialogue obligations and trust as indications that the participants are rational agents.

Construction of Dialogue and Domain Information

This chapter digs deeper into one particular aspect that is essential for CDM: the process of planning and producing responses that would be perceived as natural and intuitive reactions in the ongoing dialogue. The starting point is the information that is exchanged in contributions to the dialogue, and the basic unit for this is the concept of NewInfo (new information), defined in terms of intonation phrases. The construction of a shared context takes place by evaluating and accommodating NewInfo with respect to one's own understanding of the dialogue situation, and providing implicit or explicit feedback of the success of this accommodation to the partner. While it is obvious that some higher-level expectations of the goal and the appropriate dialogue strategies are needed in order to guide reasoning, the basic approach in the accommodation is bottom-up: the identification of the NewInfo and its accommodation with the current dialogue situation is managed locally. The actual structure of the dialogue is recorded in the conceptual links between the pieces of NewInfo, and can be constructed by tracing the paths that show how the linking of NewInfo to the dialogue topic has manifested itself in the course of the interaction. In practical dialogue systems, the reasoning itself requires well-defined and rich semantic representation and an ontology-based reasoning engine.

Dialogue Systems

In this chapter, different dialogue system architectures, with special reference to agent-based architectures, and some basic dialogue management techniques are introduced. Suitable representations are also discussed briefly. The chapter clarifies what has been implemented and what is capable of implementation on one hand, and what is still required for further experimentation and research, on the other. The chapter attempts to establish where CDM can go beyond application development and then furnish interactive systems, taking a wider view about how spoken language interaction takes place and how it could be managed. Examples of CDM-based dialogue systems are also presented.

Constructive Information Technology

Some variations of the CDM-style dialogue management are discussed in this chapter. In particular, issues related to adaptivity and learning are investigated, as well as the notion of "full-blown communication" in the context of multimodal and non-verbal communication. This brings us back to the discussion of the general applicability of the aspects of human-human communication to human-computer interaction, and to the main claim that the genuine human factor in the design and development of intelligent interactive systems is the system's ability to use natural language communication, i.e. usable, flexible, and robust interactive systems should *afford* natural interaction.

Conclusions and Future Views

This chapter summarises the contributions of the book and presents some future challenges for CDM-based interaction management. It defines the book as a theoretically-based overview of interaction technology, which will hopefully be useful for researchers and students of interactive systems as well as for developers of commercial spoken dialogue systems.

Acknowledgements

Writing this book has been a lengthy and challenging project and would not have been possible without the help and support of many colleagues and friends. It is impossible to list all of those who have contributed to the ideas and their development and implementation in one way or another, either as my students, team members, and co-workers in research projects, or as dialogue partners and commentators in meetings and casual coffee conversations. A collective thank you must suffice to express my gratitude to all of you.

However, there are a few people I would especially like to mention since their influence has been crucial in various stages of the writing, with explicit comments and/or advice on the text.

First, I would like to mention Jens Allwood (University of Gothenburg, Sweden) whose work on Activity-based Communication Analysis has long been a source of inspiration for the CDM model. I have enjoyed our many discussions and arguments on the linguistic-philosophical aspects of communication, rational agents, communicative functions and language-based interaction, and his comments on the earlier versions of some of the chapters have been insightful.

I would also like to mention Michael McTear (University of Ulster, Northern Ireland) who has supplied me with another angle on spoken interaction. Through commenting on some of the chapters, and collaborating in various workshop organisation and writing activites, Mike has provided direct and indirect help with a more practical view of interaction, as well as useful guidance for developing advanced dialogue systems.

The late Karen Spärck-Jones (University of Cambridge, UK) should also be mentioned here. Besides providing a general model for scientific thinking, her encouragement and sharp observations have greatly shaped the structure and arguments of the book, especially in the early phase of its writing, when I was a visting Fellow at the Computer Laboratory and had the pleasure of using her "library" as my office.

I would also like to mention my husband Graham Wilcock, for his comments on many of the chapters, including his patient corrections of typos and grammar. Without his continuous support and love the whole book would have been impossible.

Finally, I would like to thank my publishers for their patience and kind understanding throughout the various stages of the project.

Kyoto

1

Introduction

1.1 Two Metaphors for Interaction Design

1.1.1 The Computer as a Tool

The world around us is changing rapidly and we are increasingly surrounded by electronic devices which provide us with various types of information. Internet and mobile applications are commonplace and future predictions describe a world where computers are embedded increasingly in the environment in which we live and in the ways in which we work: in a few years' time, for instance intelligent home robotics will cater for our many everyday needs, and we will be required to interact with a complex environment which will consist not only of people, but also of computers embedded in our daily surroundings. Technological roadmaps have envisaged how smart environments will be populated by several context-aware devices which communicate with each other and with the users, and how the future information and communication systems will contain computers built into products such as clothes, books, beds and sporting gear (Sjöberg & Backlund, 2000; Plomp et al., 2002). The systems will identify their current context of use, adapt their behaviour and allow for natural interaction. Computers will also have senses and be able to interpret human expressions, will be able to smell, feel, hear, see and taste, and there will be intuitive human-computer interfaces that mimic human communication.

In such environments, our interactions with the computer will also become more complex. The internet provides an information source as well as an infrastructure for embedded computing and virtual interaction, while mobile communication and wireless appliances can be deployed for building various digital services. Many of our everyday tasks already require the use of a computer as a tool: text editing, image processing, communication with friends and colleagues by emails and web-phones, information seeking in the internet and in various digital databases. Computers also enter into human activities where they replace the other participant: interactive situations with automatic services are commonplace, and different types of online services (news, banking, shopping, hotel and flight

Constructive Dialogue Modelling   Kristiina Jokinen
© 2009 John Wiley & Sons, Ltd

reservations, navigation, e-government and e-administration) are in extensive use. Moreover, computer agents can also act on behalf of the user when given the user's identity in simulated worlds, but also in the real world as e.g. taking care of automatic payments or news watching when assigned the user's requirements and preferences.

Often these interactions presuppose communication in some form of natural language: information seeking and question answering, route navigation and way finding, tutoring, entertainment and games, dialogues with robots and talking heads, not to mention human-like conversational settings such as negotiations, monitoring of meetings and social chatting all involve communication using natural language words. Moreover, interactions with computers are not necessarily conducted through graphical interfaces with the keyboard and mouse, but can also include multimodal aspects such as speech, maps, figures, videos, and gesturing, and the screen need not only include text boxes and windows but also animated agents and talking heads which create the illusion of a fellow agent.

So far the general view of human computer interaction has regarded the computer as a tool: it supports human goals for clearly defined tasks (see e.g. Shneiderman, 1998). Interface design has focused on designing easy-to-use artefacts, and the main research activities are concerned with usability issues, i.e. how to balance task requirements and human factors so as to assist the user to perform the intended task as efficiently as possible. Main applications have dealt with graphical user interfaces (GUI) and web portals, as well as assistive tools for such complex tasks as computer-aided design and computer supported collaborative work where the interface is intended to support the users' work and thinking. The emphasis on the interface design has been on clarity, simplicity, predictability and consistency, which make the interface easy and transparent to use. Moreover, the user should have control of the properties of the interaction, i.e. of the system's adaptive capabilities.

The same principles have also been applied to the design of natural language interfaces. Speech is often regarded as an alternative mode that can be fixed on top of GUI-type interfaces in order to read out menu commands or to give a more natural sound to dialogue type interfaces. The main emphasis is thus on system responses which should be geared towards clear, unambiguous prompts. The prompts should take the user's cognitive limitations into account and provide the user with explicit information about the task and the system's capabilities (see e.g. Dix et al., 1998; Weinschenk & Barker, 2000). In the development of speech-based applications, the emphasis has also been on a high degree of accuracy of speech recognition and word-spotting techniques so as to pick the important concepts from the user input. The so-called How May I Help You (HMIHY) technology (Gorin, Riccardi & Wright, 1997) has been influential in this respect, and helped to build applications where, typically, the user can reach her goal after a few spoken exchanges. To minimize the speech recognition errors, the user is given only a couple of choices to choose the response from, and the interaction is strictly system-directed, i.e. the dialogues are driven by the system questions.

Speech has often been regarded as an alternative mode that can be fixed on top of GUI-type interfaces in order to read out menu commands or to give a more natural sound to dialogue interfaces. However, as pointed out by Yankelovich (1996), speech applications are like command line interfaces: the available commands and the limitations of the system are not readily visible, and this presents an additional burden to the user trying to familiarise herself with the system. The heart of effective interface design thus lies at the system prompt design as it helps users to produce well-formed spoken input and simultaneously to become familiar with the functionality that is available. Yankelovich herself gives best-practise guidelines in her SpeechAct system, and introduces various prompt design techniques to furnish the system with natural and helpful interface. Some such techniques are tapering, i.e. shortening the prompts as the users gain experience with the system, and incremental prompts, i.e. incorporating helpful hints or instructions in a repeated prompt, if the previous prompt was met with silence (or a timeout occurs in a graphical interface). The system utterances are thus adapted online to mirror the perceived user expertise.

Spoken language interfaces also set out expectations concerning the system's capability to understand and provide natural language expressions. For instance, Jokinen and Hurtig (2006) noticed that the same system is evaluated differently depending on the users' predisposition and prior knowledge of the system. If the users were told that the system had a graphical-tactile interface with the ability to input spoken language commands, they evaluated the spoken language communication as being significantly better than those users who thought they were interacting with a speech-based system which also had a multimodal facility. A spoken language interface appeared to bring in tacit assumptions of fluent human communication and consequently, expectations of the system's similar communication capabilities were high, making the actual experience disappointing as the expectations were not met. With the graphical-tactile interface, speech was not regarded as the primary means of communication; instead speech provided an interesting extra value to the interaction, and thus was not judged as harshly.

When dealing with the present-day interactive services, however, users need to adapt their spoken communication methods to the those available in the technology. Interactions with natural language dialogue systems are usually characterised by repetitive and unhelpful patterns which are designed to operate in a limited context on a particular task, and to provide easy-to-process input for the system. They leave the users unsatisfied if the usage context is different from that designed in the system. Moreover, the problem is not only that the users may not complete the task that the system is meant to be used for, but that the interaction itself does not show the sensitivity to the users' needs that characterise human communication. In other words, spoken dialogue systems are not only required to meet their task specification, but they should also cater for the user's needs concerning intelligent knowledge management and adaptation to the situation.

Human-human communication involves the smooth coordination of a number of knowledge sources: the characteristics and intentions of the speakers, the topic and focus of the conversation, the meaning and frequency of lexical items, multimodal aspects such as gesturing, facial expressions, and prosody, as well as communicative context, physical environment, world knowledge, etc. The challenges associated with spoken language interface design can be identified as three complex areas: natural language meaning, dialogue strategy, and user adaptation. All of them are related to knowledge management and construction of the shared context. For instance, natural language meaning does not simply function in a compositional manner whereby the semantics of a construction would be the sum of the semantics of its components. Rather, the meaning is constructed via interaction between the item and its context. It can be said that the meaning of an item is the sum of all its contexts (see the computational techniques in Lund and Burgess, 1996, and in Sahlgren, 2005), and thus the meaning of natural language is affected by the contexts in which it is used and shared. Various dialogue strategies are also important when building and managing the shared context. The speakers use natural language to clarify, explain, and describe complicated issues, and they apply their intuitive behaviour patterns, politeness codes, and communication styles to interpret and react to the partner's utterances. Finally, the speakers also adapt to each other and the shared context, and tailor the interaction according to their particular needs. Gradual alignment with the partner takes place on all levels of communication: lexical items, sentence structure, pitch level, intonation, body posture, etc. (see, e.g. the experiments on the speaker alignment in Pickering & Garrod, 2004). The three interface challenges will be discussed more in the later chapters of the book.

We must note that if the features of human-human communication are transferred to spoken interactive systems, the users are involved in an activity that differs from and even contradicts what they are used to when usually communicating with natural language: although the interaction mode is natural language, the interacting partner is a tool and not a fellow human. Since humans are extremely adaptable users, they can learn interaction patterns that are unintuitive in human communication (e.g. wait for a signal to speak, or mark with carriage return or push button to end their turn), and also learn to speak in a clear and articulated manner. However, the very use of natural language as an interface mode also directs the user's actions towards communication rather than simple manipulation of objects and task completion. The computer is treated like a fellow partner in the communicative situation, and natural language interaction seems to inherit the same kind of intuitive interaction patterns that prevail in human-human communication. For instance, the studies by Reeves and Nass (1996) show that the users anthropomorphise, i.e. assign human characteristics to the computer. The users see personality in computers, and assess the personality very quickly with minimal clues. The perceived personality further affects how the users evaluate the computer and the information it gives to the user. For instance, humans think highly of computers that praise them, even though praise is expressed as obvious flattery and may not be deserved at all. The users also rated the quality of a computer more frankly if the evaluation questionnaire was presented to them on another machine than the one they were meant to evaluate; Reeves and Nass concluded that the users unintentionally avoided hurting the feelings of a computer which they had just interacted with.

People thus seem to equate media with real life, and the computer with a human-like agent. Even though they may not realise that this is what they are doing, the complexity and unfamiliarity of interactive applications seem to give reasons to consider the computer as a communicating agent rather than a controllable tool. Indeed, the future digital environment that is envisaged in the ubiquitous computing paradigm (Weiser, 1991: http://sandbox.xerox.xom/ubicomp/) encourages such interaction: digital services, sensory devices, etc. populate our environment and penetrate into all aspects of life, ready to change our working habits and interaction patterns in a fundamental way. Digital participants are included as other communicating partners, and the appliances are aware of the other appliances as well as the users, they adapt and respond to the users' needs and habits, and are accessible ubiquitously via natural interaction.

New interaction techniques are also being developed to complement and compensate conventional graphical, mouse and keyboard interfaces. Besides speech, technologies for tactile, gesture and eye-gaze interaction have matured in the past ten years, and enable the user's multimodal communication with applications. Taking advantage of the entire repertoire of different media and communication channels, and by mimicking the possibilities available in human communication, these novel techniques aim at supporting richer and more natural interactions with applications. The term *Perceptual User Interfaces (PUI)* has been introduced (Turk & Robertson, 2000) to cover interfaces that combine natural human capabilities (communication, motor, cognitive, perceptual skills) with computer I/O devices, and thus include multimodal communication in the wide sense. Although perception and multimodal communication is natural and intuitive in human communication, many open issues still deal with the design and use of such systems. Especially, a deeper understanding is needed of the manifestations and functionality of the information received via different input channels, and of the relation between verbal and non-verbal communication for the exchange of messages: how do human speech and gestures, gaze and facial expressions contribute to communication, how are the presentation modes and the symbolic content coordinated, interpreted and integrated into meaningful actions? However, within the wide variety of novel interface technology, it is easier to see the system realised as a conversational interface agent that mediates interaction between the user and the application, than as a tool which is used to control and perform certain aspects of a task.

It is also important to survey tasks and applications where communicative systems would be superior to directly controllable systems. For instance, conversing toasters may not be as impressive showcases for spoken dialogue systems as hand-held conversational devices that inform the user about appointments or tourist attractions. Tasks with a straightforward structure may not allow the benefits of flexible natural language conversation to come through: although call routing and voice controlled applications are useful speech-based services, their interaction capabilities need not go past question answering dialogues for which present-day interaction technology already provides management techniques. Instead, tasks that require complex reasoning, evaluation, and integration of pieces of information seem more appropriate for rich dialogue interactions. Indeed, viable spoken dialogue application domains in this respect have also appeared: guiding and training novice users for skill acquisition in tutoring systems, navigation and way finding for location-based services, speaking in games and entertainment, social chatting in human-robot interactions. Besides information search and presentation, these tasks presuppose the participants' ranking of their immediate goals, linking of focussed information to their previous knowledge, and balancing of long-term intentions with the partner's requests. Interactions are not necessarily predictable in advance but depend on the context and actions of the individual participants: communicative situations are constructed dynamically through the participants' communicative (and physical) actions. The driving force for interactions thus appears to reside in the participants' intention to exchange information, while the flow of information is controlled by the participants' awareness of the situation and by their reactions to the contextual changes. Consequently, the tasks require communication of certain intentions and knowledge, negotiation skills, and awaress of the partner's needs, and there appears to be a need for a user-friendly flexible interface that would understand the logic behind natural language utterances as well as that of communicative principles.

1.1.2   The Computer as an Agent

In the past years, a new metaphor for human-computer interactions has thus emerged: the computer as an agent. The computer is seen as an assistant which provides services to the user, such as helping the user to complete a task by providing necessary information of the different steps of the task (e.g. tutoring, explaining, medical assistance), or helping the user to organize information (e.g. electronic personal assistants).

Future interactions with the computer are envisaged to use possibilities offered by multimodal interaction technology and to resemble human communication in that they are conducted in natural language. The new types of interfaces are expected to possess near human-like conversation capabilities so as to listen, speak and gesture, as well as to address the user's emotional state in an intelligent manner.

On the other hand, human control over the system's behaviour is not so straightforward and as easily managed as before. Computers are connected with each other into networks of interacting processors, and their internal structure and operations are less transparent due to distributed processing. Object oriented programming and various machine-learning techniques make the software design less deterministic and the input-output relation less controllable. Context-aware computing strives to build systems that learn the preferences of individual users while also understanding the social context of user groups, adapting decisions accordingly. Simply, the view of the computer as a tool which is a passive and transparent "slave" under human control is disappearing.

It must be noticed, however, that the agent metaphor has been used in three different research contexts which are quite separate from each other. Consequently, the metaphor bears different connotations in these contexts. Below we will discuss the notion of "agent" in interface design, robotics, and software design. One of the obvious uses of the agent metaphor is connected to interface agents, i.e. to various animated interface characters which aim at making the interaction more natural. It is argued that personification of the interface makes interaction less anonymous, and thus contributes to the user's feeling of trust in the system. For instance, in the SmartKom project (Wahlster, 2004), the metaphor was realised in the Smartakus – an interface agent capable of mediating interaction between human users and the application: it took care of the user's requests with respect to different applications and application scenarios. Other examples, and of more human-like animated agents, include the museum guide August (Gustafson et al., 1999), real estate agent REA (Cassel et al., 2001a), Medical Advisor GRETA (Pelachaud et al., 2002a), and Ruth (DeCarlo et al., 2002).

Interface agents equipped with human-like communicative, emotional and social capabilities are usually called Embodied Conversational Agents (ECAs) (Cassell et al., 2003). ECAs have their own personality, different abilities, and they employ gestures, mimics and speech to communicate with the human user, see e.g. Pelachaud and Poggi (2002) for a discussion on multimodal embodied agents. André and Pelachaud (forthcoming) provide an overview of the development of ECAs, starting from TV-style presenters (e.g. ANANOVA, http://www.ananova.com) through virtual dialogue partners such as Smartakus, to role playing agents and multiparty conversational partners. Scenarios where human users truly interact with synthetic agents are still experimental, however. The earlier ones concern immersive virtual environments where the users can interact with the agents, and include such systems as CrossTalk (Gebhard et al., 2003) and VicTec (Paiva et al., 2004).

André and Pelachaud (forthcoming) also point out that the interaction with interface agents creates a social environment where norms and cultural rules are expected to be followed. Most ECA studies explore various social aspects of interaction and experiments are designed to bring in novel insights concerning interaction with synthetic agents. In many cases, the studies concern game

and entertainment environments: e.g. Isbister et al. (2000) studied social interaction between several humans in a video chat environment, while Becker and
Wachsmuth (2006) focus on a game scenario in which the agent takes on the role
of a co-player.

Interaction management itself is usually based on schemas. However, in the
system developed by Traum and Rickel (2002), plan-based dialogue management
techniques with dialogue acts turn-taking and grounding are used, and extended
in multiparty dialogue contexts. The task concerns rehearsal of military missions,
and the dialogues take place between the human user and virtual agent or between
virtual agents. The interaction model allows reasoning about the other agents and
their intentions, emotions, and motivations, and later developments also include
teamwork and negotiation (see e.g. Traum et al. 2003). The agents are thus capable
of simulating social interaction, and interesting research topics in this respect
concern e.g. cultural and personality differences. Also in the Collagen system
(Sidner, Boettner & Rich, 2000), the interface agent is a collaborative agent which
implements the SharedPlan discourse theory (Grosz & Sidner, 1990, Grosz &
Kraus 1995) and plan recognition using hierarchical task network (Lesh, Rich &
Sidner 1998). The framework has also been integrated in animated agents (Cassell
et al. 2001).

Another use of the agent metaphor is found in AI-based interaction management. The BDI (Belief-Desire-Intention) agents refer to autonomous intelligent
systems that model such cognitive skills as planning and reasoning so as to be
able to deliberate on the appropriate action in a given situation. For instance, the
last two examples above can also be said to belong to this tradition as they use
the classic techniques developed in the context of communicating BDI-agents.
A continuation of this research can be found in robotics, although probabilistic
and machine-learning techniques have mostly replaced the traditional handcrafted
reasoning rules, and novel applications range from intelligent cars to soccer playing robots. Autonomous robots are also called Situated Embodied Agents (SEAs)
as they are directly interacting with their environment: the robots are equipped
with various sensors and vision components through which they observe their
environment and with motor components through which they produce suitable
actions. Interaction with the environment is based on rapid reasoning about the
appropriate action in a given situation and, if the robots work as a team, on
communication with one another, so as to coordinate their collaboration.

If the robots are placed in situations where the task requires human intervention, communication is most naturally conducted in natural language. One
of the first examples of such situated dialogues is the robot Godot (Bos et al.,
2003), which interacts with the user in a museum environment. Dialogues with
mobile and sensing robots also offer a novel research topic for interaction management: language grounding. Although the term grounding is well known in
dialogue research, dealing with feedback and construction of a shared understanding as part of successful communication, the challenging aspect with robots is the

opportunity to talk about the physical environment which is shared between the user and the robot. Reference to elements in the environment must be grounded taking into account the participants' vision field and their focus of attention at a given moment, i.e. a relation between the element and its linguistic expression must be established. Kruijff et al. (2007) present an ontology-based approach to the problem in the context of human-robot interaction where a human user guides the robot around a new environment, and the robot incrementally builds up a geometric map of the environment through its interaction with the user. The situated dialogue understanding is established by linking semantic representations of the utterance meaning with the robot's internal world representation. The three-layer world model connects the spatial organisation of an environment with a commonsense conceptual description of the environment, and thus provides the basis for grounding linguistic meanings into the innate conceptual knowledge of the objects encountered by the robot in its environment.

Like ECAs, robots also provide a testing ground for social interaction and expressive face-to-face communication. One of the first to engage people in natural and affective, social interaction was Kismet (Breazeal, 2002), a robotic head which was inspired by a caretaker-infant interaction to explore various social cues and skills that are important in socially situated learning. The communication robot Fritz (Bennewitz et al., 2007) is a humanoid robot, which mimics the way humans interact with each other, and thus aims to give insights of efficient and robust multimodal human communication strategies and their transference to human-machine interface. Fritz uses speech, facial expressions, eye-gaze, and gestures, and it can generate synchronised gestures and speech as well as change its attention between the communicating partners in the vicinity. The results of its public demonstrations showed that people found the robot human-like and an enjoyable communication partner. A completely radical approach to robotics is taken by Ishiguro (2005), who aims at establishing Android Science, by building robots that not only act but also look like humans too. The use of life-size and life-like android robots to improve the quality and naturalness of communication is, however, somewhat controversial, and empirical results of the subjects' perception seem to confirm the "Uncanny Valley" hypothesis. The hypothesis says that the human users first feel increasingly positive and emphatic towards humanoid robots, but at some point when the robot's appearance and behaviour becomes "almost human", the emotional response becomes negative. If the development towards more human-like robots continues, the emotional response becomes positive again.

Software agents can be classified into different types depending on their capabilities and internal processing. For instance, Russel and Norwig (2003) define the classes of agents as follows: the reflex agents respond immediately to environmental percepts, the goal-based agents act so as to achieve particular goal(s) and utility-based agents try to maximise their own "happiness".

Yet the third use of the agent metaphor is found in software engineering: that of software agents. Software agents are computational means to design and implement software, and should not be confused with ECAs or BDI-agents. They are not autonomous or intelligent agents but software constructs which "live" in system architectures. Originally as a complementary metaphor to direct manipulation interfaces (cf. discussion in Maes & Shneiderman, 1997), they can be used to automatize some tasks and services for the user, by learning to organise and filter information according to the user's personalised needs (thus they help the user to cope with the increased amount of information and maintain a close connection to interface agents discussed above). The agents can also interact with each other and their interaction can be described using the same metaphor as communication, i.e. "speech acts", defining the pieces of information being exchanged (cf. the FIPA Agent Communication specifications). Software agents can also be classified into different types depending on their capabilities and internal processing. For instance, Russel and Norwig (1995) define the agent widely as anything that can perceive its environment and act upon that environment, and then consider the types of programs that the agents can implement to map their percepts to actions, as follows: the reflex agents respond immediately to environmental percepts, the goal-based agents act so as to achieve particular goal(s), and utility-based agents try to maximize their own "happiness".

The agent-based system architectures usually refer to object-oriented programming frameworks within which the application design is implemented. In this case the software agents are pieces of software that encode specific methods to accomplish a task, and they can be grouped into software libraries that will also be available for other applications. The benefits of software agents concern distributed and asynchronous processing, which provide freedom from a tight pipeline processing and can speed up the system performance as well as support a more flexible functioning of a complex system. The need for decomposing the main task into smaller indpendent subtasks usually also helps to resolve the main task more easily than if done in a single block computation. A distributed architecture can also allow experimentation with different computational models of the task and consequently, support flexibility on the level of system design and architecture. However, the design of practical applications using software agents is a skill requiring specialised knowledge: some reference architectures exist (see e.g. Jokinen and McTear, forthcoming), but on the concrete level of designing the system, the task and user requirements are varied and depend on the particular task, domain and interaction management goals.

Progress in various research areas related to building intelligent interactive systems has been remarkable in recent years, and both engineering and human communication studies have yielded important insights in interaction and context management. Developments in speech recognition and synthesis, natural language processing, computational methods and techniques, architectures and hardware design have made it possible to build interactive systems that can be applied in real-world tasks. Despite the advances in technology, the natural interaction mode in various interactive applications is mouse, menu and graphics, rather than natural language. However, dialogue capabilities allow the system to function in a wider range of situations, and provide services in an easier and more natural manner.

One of the early advocates of intelligent dialogue interfaces was Negroponte (1970) who, argued in favour of a rich interaction language for designer-machine interaction. The starting point was to remove barriers between architects and computing machines, and to enable computer-aided design to have machine intelligence that would allow cooperative, natural interaction. A rich interaction language would bridge the gap between a designer who wants to do something and the execution of the design: the designer need not be a specialist "knobs and dials" person but could formulate the problem and specify goals in his or her idiomatic language, while the design machine could respond, in the same language, concerning the constraints and possibilities of the design.

Most recently, Sadek (1999) pointed out that it is the system's intelligence that provides the good interface and ergonomy of the service. He argued in favour of rational dialogue agents, i.e. a design paradigm for practical systems based on formal modelling of rational agents and their reasoning. Machine intelligence, as the fundamental goal for natural interaction, is possible via a flexible framework that allows rationality, communication and cooperation to be taken into account in the functional requirements of the system, and supports reasoning about mental attitudes and communicative acts.

However, creating a fully-fledged conversational agent is a highly demanding task. As argued in Jokinen (2000) dialogue systems should be investigated as learning systems rather than static models of service providers: the intelligence of a dialogue system depends on the available interaction patterns as well as on the factual and conversational knowledge that the system can deploy in its rational reasoning. In order to equip dialogue systems with the appropriate knowledge, it is not sufficient, or even possible to store the world knowledge explicitly in the computer memory (although with all the information in the web and digital libraries this now seems more likely than a couple of years ago), but it is necessary to include reasoning and dynamic reasoning and update procedures in the system so as to enable the system to use its knowledge and to learn and adapt itself to new environments and to new interaction patterns. For intelligent interaction, it is thus necessary to study how the agents use their full conversational possibilities, including verbal and non-verbal interaction strategies, recognition of new and old information and information presentation via different modalities.

1.2   Design Models for Interactive Systems

The system design needs to address the following questions:

1. What should the system do?
2. How should the system do what it is meant to do?

The first question leads to the design of system functionality, including the design of the system's interaction model. The second question is directed towards architectural and processing issues: when the models of the system functionality, the task and the required interaction capabilities are available, the next step is to decide how to implement the models.

In this section we discuss solutions to the second question, loosely following the agent classification presented by Russel and Norwig (2003), see Section 1.1.2, and focusing especially on the consequences the different types of agents have on the interaction management between the user and the system.

1.2.1 Reactive Interaction

In the simplest case, human-computer interaction can be enabled by the direct mapping of the user input onto a certain system reaction. Interaction is thus managed by specifying the relation between input-output pairs, and reduced to a automatic reaction to a particular input. Figure 1.1 below represents this kind of situation.

Reactive systems are good for tasks where the desired actions are not too numerous to list and the input conditions can be clearly defined. Typical computer interactions are of this type: by giving a command, the user will enable a certain task to be done, and as an answer to the command receive requested information or a message confirming that the requested action has been taken. Many practical applications and speech-based interactive systems are also based on this type of interaction: for instance, banking cash machines and traffic information providers, robot vacuum cleaners and movement detectors. In reactive systems, interaction management is included in the interface design, and there is no explicit model of the interaction: this is implicit in the system functionality. Interaction is managed with the help of scripts which describe the various input possibilities open for the user, and the outputs that the system should provide as a reaction to each input. This can be implemented as a finite-state machine, with deterministic state transitions defining input conditions and output reaction. The rationality of the interaction is based on the designer's understanding of what kind of functionality is needed for successful task completion and for the application to operate smoothly, but the system itself cannot reason e.g. what the inputs are good for or why the actions are executed in a

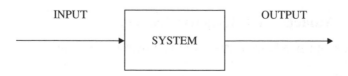

Figure 1.1   Straightforward reaction

certain order. On the other hand, in many cases reactive systems provide a reliable and sufficient way of managing simple interactions, and in some cases, like taking care of frequently occurring dialogue phenomena, they can speed up the system performance, and also provide some analogue for human reactive behaviour.

1.2.2   Interaction Modelling

If the task becomes more complicated and requires an elaborated combination of the input and output conditions, reactive systems will rapidly become cumbersome. The task may also require that some pieces of information must be known before a certain action can be taken, but the manner and order in which this information is obtained is unimportant: the information may be supplied by the user in one go, or the system may acquire it piece-meal through internal actions. For instance, in travel service systems, a user may have asked for travel information by giving information about the departure and arrival locations but no departure time ("I want to go from Eira to Pasila"), while another user may ask a similar question by giving information about the departure location and departure time but no arrival location ("I want to leave from Eira at about 11 am"). Although it is possible to identify the alternatives under different input conditions, this solution fails to make the generalisation that in both cases the user has given information about the departure location, while other necessary information is still to be found out. Moreover, when the amount of missing information increases, possible parameter combinations also increase, and the clear structure of input-output relations starts to fall apart.

In order to manage this kind of interaction in a flexible way, we need to separate the knowledge about the task from the knowledge about the interaction. Although interaction and task management are closely intertwined, they are conceptually separate models: the task model describes the agent's knowledge of what is needed to complete a particular task, while the interaction model describes procedural knowledge of how to achieve the goal through interacting with the partner. For instance, in order to complete the task of retrieving travel information, it is necessary to know the departure place, the arrival place and the departure or arrival time, and if any of the necessary information is missing, the agent needs to formulate a query, and enter into interaction with the user on that particular point. Task model is often simply a list of task-related concepts that the system needs to talk about, although hierarchical task structures and elementary reasoning on subtasks and possible alternatives can be included (cf. expert systems in the 1980's and recent developments in the semantic web). The interaction itself can be understood in different ways, and has been described as straightforward question-answering, negotiation, information updating, plan execution, collaboration on a particular task, decision making, or, as in this book, construction of a shared context in which the agents act. Figure 1.2 depicts this kind of systems

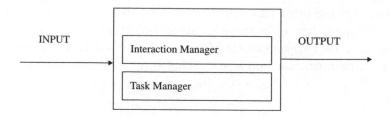

Figure 1.2 Interaction management

with two separate components (managers) taking care of the processing of the two types of knowledge.

In dialogue management, this kind of interaction can be enabled by various techniques where the system maintains a model of the task in the form of a plan or a frame, and operates on the knowledge either in a procedural way as part of the implementation of the whole system, or in a more transparent way by maintaining a separate interaction model where possible dialogue acts, moves and updates are described in a declarative way (see different modelling approaches in Section 2.2 and dialogue management issues in Section 2.3).

1.2.3 Communication Modelling

Although the system's flexibility can be improved with a separate dialogue model component, the computer agents are assumed to operate in a world that is accessible and deterministic: the knowledge is complete for the purposes of the task that the agents want to perform. The designer of the system is an omniscient creator of a virtual world, where the two participants, i.e. the application and the user, interact under logically closed knowledge conditions. The interactive situation can be depicted as in Figure 1.3.

However, the closed-world approach does not lend itself to situations where the partners' knowledge is too large and varied to be enumerated by the designer in advance. For instance, in various ubiquitous computing situations it may be impossible to know and manage all the information or even all the relevant information for a given task, since the agents and the world constantly change. The knowledge is also inherently limited to the particular view-point of the agent at a given moment and the agents' deliberation of the suitable actions is based on heuristics rather than logical inference (rational agents exercise "minimal rationality", cf. Cherniak, 1986). The agents do not necessarily act so as to follow the partner's wishes benevolently either; quite the contrary, the agents can have conflicting goals, and they need to negotiate about the alternative goals. Moreover, interaction does not only take place between two partners but can include a group where communication follows group dynamics rather than individual dialogue strategies.

The interaction between agents in constraint rationality conditions resembles the situation in Figure 1.4. The two agents have their own knowledge of the world

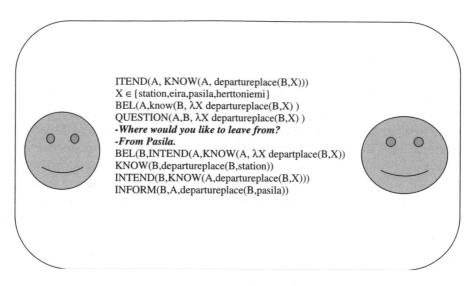

ITEND(A, KNOW(A, departureplace(B,X)))
X ∈ {station,eira,pasila,herttoniemi}
BEL(A,know(B, λX departureplace(B,X))
QUESTION(A,B, λX departureplace(B,X))
-Where would you like to leave from?
-From Pasila.
BEL(B,INTEND(A,KNOW(A, λX departplace(B,X))
KNOW(B,departureplace(B,station))
INTEND(B,KNOW(A,departureplace(B,X)))
INFORM(B,A,departureplace(B,pasila))

Figure 1.3 Omniscient view of dialogue management

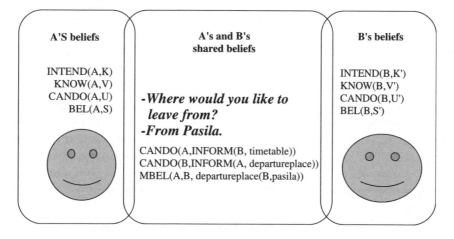

A'S beliefs

INTEND(A,K)
KNOW(A,V)
CANDO(A,U)
BEL(A,S)

**A's and B's
shared beliefs**

*-Where would you like to
leave from?*
-From Pasila.

CANDO(A,INFORM(B, timetable))
CANDO(B,INFORM(A, departureplace))
MBEL(A,B, departureplace(B,pasila))

B's beliefs

INTEND(B,K')
KNOW(B,V')
CANDO(B,U')
BEL(B,S')

Figure 1.4 Constraint Rationality

and the communicative situation, as well as their own intentions and motivation
for the dialogue. Their actions are based on their deliberation with respect to their
private beliefs and intentions, and what has been gathered during the interaction
as mutual beliefs. Mutual beliefs are part of the shared knowledge, constructed
during the dialogue through the presentation and acceptance cycle (Clark, 1989).
The model does not assume an omniscient designer but rather, the necessary world
is constructed in the course of the interaction.

Traditionally dialogue management models follow the logical structure of the
interaction management task: input analysis, dialogue handling, output generation

(see Bernsen et al., 1998). This supports pipelined architectures where the system components operate in a fixed order, and the dialogue manager is seen as one single module in the overall cycle of interaction. More adaptable dialogue management models became available when object-oriented programming started to gain popularity. Agent-based architectures provide flexibility and asynchronous processing so that the components can, in principle, operate independently yet in an integrated manner on the dialogue data. Of course, the logical order of the management task must also be respected, so in practise, the components tend to work in a particular order. However, the order is not meant to be imposed by chaining the software agents to follow the interaction cycle but by making the agents react to follow a particular occurrence pattern in the information that is available in the system: the agents can react to particular information states only, while their reactions can change the current state of system so that it becomes suitable for another agent to react. The order is thus a side effect of the available information at a particular time in the system as a whole.

The script- and frame-based dialogue management models are not well suited to take care of the communication between agents that interact in constraint rationality situations. In these situations, dialogue management requires special modules for the agents' intention and belief reasoning, as well as efficient coordination of various modalities and multiple knowledge sources, and the basic versions of these models do not offer rich and flexible enough tools for this. Communication between agents in constraint rationality situations requires special modules for the agents' intention and belief reasoning, as well as efficient coordination of various modalities and multiple knowledge sources. Consequently, dialogue management requires rich and flexible enough tools for this. Traditionally dialogue management has followed the logical structure of the interaction management task: input analysis, dialogue handling, and output generation (see Bernsen et al., 1998). This supports pipelined architectures where the system components operate in a fixed order, and the dialogue manager is one single module in the overall architecture.

More adaptable dialogue management became available when object-oriented programming started to gain popularity. Agent-based architectures provide flexibility and asynchronous processing, so that the components can, in principle, operate independently yet in an integrated manner on the dialogue data (see some discussion e.g. in Blaylock, Allen & Ferguson, 2002; Kerminen & Jokinen, 2003). Of course, the logical order of the management task must also be respected, so in practise the components tend to work in a particular order. However, the order is not meant to be imposed by chaining the software agents to follow the interaction cycle but by making the agents react to a particular occurence pattern in the information that is available in the system: the agents can react to particular information states only, while their reactions can change the current state of system so that it becomes suitable for another agent to react. The order is thus a side effect of the available information at a particular time in the system as a whole.

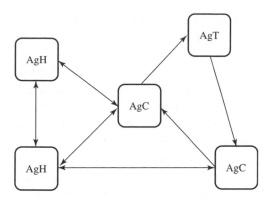

Figure 1.5 Schematic depiction of the interface in Agent-based dialogue management

Agent-based dialogue management thus seems to offer a framework for responsive and constructive dialogue management, especially in the context of ubiquitous communication environment where an increasing number of interactive intelligent applications occur and require the users' attention. Figure 1.5. gives a schematic depiction of the interface in this kind of interaction: it shows the communication modelling approach with different actors interacting, cooperating and coordinating their actions with each other (some of the participants are humans (AgH), some intelligent computer agents (AgC), while others (AgT) do not necessarily have any characteristics of agency or constraint rationality, but are controlled by humans and the more autonomous agents).

1.3   Human Aspects in Dialogue System Design

The emphasis in this book is on natural communication, which is regarded as the main source for intelligent human-computer interaction. There are also many other disciplines that deal with the human aspects of interaction modelling and, the main approaches are discussed briefly in this section. The different departure points cause variations in the primary research topics and methods, but they all share the main goal of designing usable systems and useful artefacts for practical tasks. The discussion leads to the conclusion of this chapter, namely that in designing interactive systems that address the requirements for intelligent interaction within the ubiquitous computing paradigm, the most important human factor is, in fact, the capability for natural intuitive and cooperative communication.

1.3.1   User-centred Design

User-centred design aims at designing end-products that are usable in the specified context of use. The approach emphasises the role of the user in the design cycle, and derives its name from the involvement of the users in the design and testing of the prototypes. The users' are taken into account early in the design cycle in order to provide information on the desired features and influence the system development in the early stages. Evaluation typically focuses on the quality of interaction between the users and the products, i.e. on usability. The product development proceeds through iterative refinements whereby the users' feedback is used to modify and further elaborate the initial design (Norman & Draper, 1986; Daly-Jones et al., 1999). The users' tasks and the environment need to be understood, too, thus bringing into focus a wide context of organisational, technical and physical factors. Computer applications are seen as end-products which should be effective, efficient and satisfying to use. Special attention is also paid to the understanding and specification of the context of use. The application should be designed with reference to the characteristics of the intended users such as their knowledge, skill, experience, habits and motor-sensory capabilities. The usability of the product can be measured with heuristics that describe the fitness of the product with respect to a set of heuristic parameters (Norman, 1988; Nielsen, 1994).

1.3.2 Ergonomics

Ergonomics looks at the design process from the point of view of human anatomy, physiology and psychology. Its goal is to apply scientific information to the design of objects, systems and environment so that the user's capabilities and limitations are taken into account. For instance, Wickens and Holland (2000) discuss how human perception and cognitive processing works, and also provide various examples of how cognitive psychology can be taken into account in designing automated systems.

In human-computer interaction, much research has focused on ergonomy. Starting from the design of the computer hardware (keyboard, screen, mouse, etc.), ergonomics is also applied to interface and software design so that they would support easy manipulation and transparent functionality of the system.

An important concept in this respect is the cognitive load. It refers to the demands that are placed on a person's memory by the task they are performing and by the situation they find themselves in. It depends on the type of sensory information, amount of information that needs to be remembered, time and communication limits, language and other simultaneous thinking processes. Several investigations have been conducted on automatically detecting the symptoms of cognitive load and interpreting them with respect to the user's behaviour (e.g. Berthold & Jameson, 1999: Mueller et al., 2001). However, it is difficult to operationalise cognitive load and provide guidelines for its measurement. It is obvious that cognitive load increases if the user's attention needs to be split between several information sources, but the different types of information, e.g. visual and spoken, can also effectively support each other. For instance, when studying different properties of presentation styles in a mobile device, Kray et al. (2003) noticed that cognitive load for textual and spoken presentation is low, but when the more complicated visual information is used, or more complicated skills are needed to master the interface, cognitive load increases rapidly and the interface becomes uncomfortable.

The work on interface design has focused especially on system prompts which should elicit required user input successfully as an answer to the system question. Best-practice design principles emphasise clear, unambiguous prompts which provide the user with explicit information about the task and the system's capabilities. For instance, the 20 Laws of Interface Design, proposed by Weinschenk and Barker (2000) introduce such aspects as linguistic clarity, simplicity, predictability, accuracy, suitable tempo, consistency, precision, forgiveness and responsiveness, in order to make the interface easy and transparent to use. Another important feature of the interface is to help the user to feel in control of the application: it is a tool that the users use in order to accomplish a certain task, and to encourage positive experience, the users should feel that they master the usage of the tool (Shneiderman, 1998). Yankelovich (1996) pointed out that the systems should also prompt the users with appropriate help which guides the users if they are having problems. She suggested different design strategies to cope with different users so that the interaction appears to be helpful. For instance, her own SpeechActs design builds on the requirements of how to design system prompts that show the development of the interaction and address problems with the interaction.

1.3.3   User Modelling

As discussed in Section 1.2.3, the aim of conversational dialogue management is to give means for flexible and rational human-agent interaction by studying human communicative capabilities and building models for their computational treatment. Research has focused on such aspects as the speaker's intentions, speech and language errors, response planning, mutual beliefs, shared information and rational cooperation. Individual users are often taken into account by constructing special user models, explicit representations of the properties of a particular user. Traditionally user models have focussed on the user's beliefs and intentions so as to support knowledge intensive reasoning (McCoy, 1988; Paris, 1988; Chin 1989) but in the practical dialogue systems, they often are simple lists of user preferences (for an overview, see: Kobsa & Wahlster, 1989; Fischer 2001).

User models are usually associated with system adaptation. For example, Jokinen et al. (2002) distinguish static adaptation, which refers to the options that the users can choose from when making decisions on interface aspects such as colour or sound, and dynamic adaptation, which refers to on-line adaptation through interaction, e.g. clustering of users on the basis of their similar navigation choices. Static adaptive features can be listed in personal profile files, but dynamic adaptation requires that the system is capable of observing user behaviour, and can learn from it. For instance, the user's familiarity with the system functionality can be traced on-line, and the system responses adapted to the suitable competence level

(Jokinen, 2005b; Jokinen & Kanto, 2004). A realistic on-line adaptation would further require dynamic updates in the system's knowledge-base, i.e. knowledge to be modelled as clusters of pieces of information that can change according to the interactions with the user.

Concerning computer agents, two opposite approaches have been prevalent in developing user models for them (Fischer, 2001): one with the goal of endowing computers with human-like abilities (human emulation), and the other where the asymmetry between humans and computers is exploited to develop new possibilities for interaction (human complementation). Since human emulation as such has turned out to be a hard task, user modelling has focussed on the complementing approach and sought for more tractable solutions in good design practises and experimental research on the user's interaction strategies. The fundamental asymmetry between humans and computer agents, however, seems to become less clear as the research and development in various kinds of interface agents, robot applications, and industrial interaction systems advances: computer agents are more likely to be assigned properties that relate their functioning to human rational properties such as cooperation, adaptation, learning. Consequently, user modelling becomes similar to intelligent interaction modelling, and the design of dialogue systems can be approached from user simulations. The latter approach has effectively been taken e.g. in Schatzmann et al., (2005), where the goal of the construction of (statistical) user models is to provide a dialogue partner for the dialogue manager during the manager's learning phase of dialogue strategies.

1.3.4   Communication as a Human Factor

The different approaches all concern human factors and aim at building systems that would to do the "right" thing at the "right" time in the "right" way (Fischer, 2001). Because of their historical roots, the approaches have focused on different aspects of the design process, and have resulted in different practical activities. Nevertheless, all of them emphasise the importance of human factors in the design and development of applications, and place demands on the system functionality from the point of view of human needs, constraints, and enablements.

When assessing the usability of final products and interactive applications, the criteria are related to the system's usefulness in the context, its naturalness, adaptation, and ability to engage the user in the communicative situation. These are also typical properties assigned to human cooperative and rational communication, and in the following chapters we will discuss how these aspects affect the user's perception of the system as an intelligent and cooperative dialogue partner, and thus her satisfaction and trust in the system as a useful and enjoyable information providing agent.

In intelligent interface design, the main human factor thus seems to be the system's ability to communicate, especially in such a way that the user (and the system) can pursue their goals effortlessly.

The main contribution of the book is to emphasise natural communication as a starting point for the interaction research. The use of natural language can be regarded as the ultimate human factor in interactive interface design, and include rationality and cooperative principles in the system architecture as part of the standard models for intelligent interaction management. Support for this is found in the ubiquitous computing paradigm where the human user is one member of a large network of communicating agents: the interaction may be of different types and of different levels of competence, but the same conversational principles apply to each agent. In the intelligent environment, where the agents need to coordinate their tasks, desires, intentions, and social activities, communication can easily become abstract and require communicative capabilities that presuppose natural language-based communication between rational and cooperative agents. In fact, it will be argued that the most affordable (Norman, 1988) interactive interface is based on natural cooperative communication.

2

Dialogue Models

It is useful to distinguish dialogue modelling and dialogue management, although they often refer to the same type of activity describing how interaction takes place. However, dialogue management refers to the computational manipulation of a particular dialogue model, and the dialogue manager is a software component that implements the designed model. Hence, dialogue management deals with the techniques, architectures and algorithms that enable dialogue technology to be developed in practical dialogue systems, whereas dialogue modelling describes the data that the dialogue manager uses to compute suitable responses.

Communication can be conducted using speech, signs and text. We refer to spoken and signed communication as *dialogue* and text-based communication as *discourse*. The latter term is also widely used to cover spoken and written communication, but we use it in a more restricted sense, and adopt *interaction* to cover exchange of information in general. This is also neutral with respect to the modality through which the information is mediated, and thus covers types of modalities other than text and speech as well, i.e. gestures, facial expressions, eye-gazing, etc.

Dialogue Modelling is an area of study where the properties, structure and processing of interaction are in focus. It seeks to advance our understanding of the principles that govern interactive situations through empirical observations of dialogues and by modelling the insights of the processes that underlie typical human interactions with computational modelling techniques.

This book is motivated by the desire to understand, elaborate and further develop the computational models of interaction which have been presented in previous research. It is thus important to set out the context and survey the field of dialogue modelling research. We start with a brief history of the field, and then discuss different approaches to model interaction, and different ways to implement intraction control in dialogue systems.

The history is not an exhaustive list of dialogue systems, but rather it aims to show how dialogue systems have developed gradually and incorporated more complicated knowledge and reasoning in the system design. Starting from counting and pattern matching systems of the 1960s, systems with limited domain knowledge

Constructive Dialogue Modelling   Kristiina Jokinen
© 2009 John Wiley & Sons, Ltd

were developed in the 1980s, resulting in a dialogue technology emerging through the large projects in the 1990s and commercial applications in the beginning of the new millennium. With further development of the technological environment, and fully-fledged communicative systems, it can be expected that research and development continues towards interactive systems operating in multiparty situations, and exploiting hybrid models for intelligent communication modelling.

2.1 Brief History

Intelligent human-computer interaction has fascinated the human mind since modern computing machines were developed in the 1940s. The initial ideas concerned systems with human-like communicative capabilities, but practical work soon refined these ideas down to more modest interaction models. However, with developments in language and interaction technology, the Internet, grid-technology and the ubiquitous computing paradigm, the old AI dream of an intelligent computer agent has appeared again, in the form of personal assistants, companions, chat-bots, robotic pet animals, humanoid robots, etc. The goal of dialogue research, i.e. to investigate and increase our knowledge of interaction, is thus even more relevant than ever before: as discussed in Section 1, in the context of intelligent computer agents it is necessary to furnish the computer with rich and natural communicative possibilities.

The history of dialogue modelling and interaction management can be divided roughly into four eras. The ideas of intelligent interaction were first discussed in conjunction with the progress of computer and Artificial Intelligence research between 1950–1970, but rapid development started only in the 1980s with the growing number of experimental prototypes that modelled various linguistically and computationally interesting aspects of interaction. This led to advances within large-scale dialogue projects in the 1990s, while the new millennium saw research mature enough to establish dialogue technology and build commercial interactive applications. Of course, research on dialogue modelling continues, and has been extended towards fully-fledged communication where non-verbal signals and multimodality are important aspects of utterance analysis and presentation.

2.1.1   Early Ideas

The early ideas of intelligent interaction can be traced back to Alan Turing who over half a century ago sketched the "Thinking Machine" in an influential article (Turing, 1950). This was an abstract machine which could engage in natural language conversations with humans and produce intelligent behaviour that could be deemed to be intelligent. Turing proposed an operational test for intelligence: a machine could be regarded as intelligent if a human judge would be unable to distinguish whether the partner was a machine or another human. The imitation game has become to be known as the Turing test, and is one of the most well-known and debated tests for computer intelligence.

Arguments in favour and against the test can already be found in Turing's article. Later discussions formulated logic-philosophical arguments concerning the nature of intelligence and how, if at all, intelligent behaviour in computers would differ from the one exhibited by humans. The debate was fostered by excitement over the novel computational tool, and the two opposite positions concerned whether the goal in AI was to make machines be like humans with human intellectual capabilities (strong AI), or whether to equip them only with the equivalent of human intellectual capabilities (weak AI).

For instance, the Physical Symbol System Hypothesis, formulated by Newell and Simon (1976), states that a well-designed and well-organised physical symbol system can exhibit intelligence of human actions: since human thinking, which by definition is intelligent, can be regarded as symbol manipulation and machines can simulate human problem solving skills by performing similar symbol manipulation, it follows that the machines can be intelligent. On the other hand, Dreyfus (1979) strongly criticises the optimism of the early AI and the basic assumptions that computers can emulate human thinking, and McCarthy (1979) considers the problems in the machine's learning and creativity and the unsuccessful attempts to programme originality into the machine thinking. Searle (1980), who invented the term "strong AI", points out that mere symbol manipulation cannot be regarded as true understanding: what is needed more is an understanding of the semantic content or the meaning that the symbols convey. Recently, the Turing test has been introduced in the form of the Loebner Prize (see below Section 2.2.3), which brings in a new aspect of intelligence: intelligence may not concern so much of the syntactic and semantic understanding of the symbols as their smart pragmatic use. Since interactions do not have one single and correct format but are constructed "on-line" as the dialogue goes on, neither can the partner's intelligence be judged solely on impeccable responses but is based on the adequacy of the responses as perceived by the judge in a certain context. The whole concept of intelligence is thus largely a subjective notion and depends on both partners: on the clever strategies of one partner, and on the favourable evaluation of one partner's behaviour by the other.

The early discussions in Artificial Intelligence cannot, of course, be regarded as dialogue modelling proper. However, it is worth noticing that intelligent behaviour is inherently tied to verbal communication in these early disputations: besides intentionality and consciousness, the capability to express oneself in words and form relevant verbal contributions is understood as one of the main signs of intelligence. The modern view of AI tends to consider the dichotomy between strong and weak AI less useful in practice, although the concept of singularity has brought in the issues for wider discussion again. The general view, however, integrates various sub-fields under the concept of the intelligent agent, and defines AI as "the study of agents that receive percepts from the environment and perform actions" (Russel & Norwig 2003). Thus the agent's communication with the environment, either via actions or language, exhibits intelligent behaviour through which the agent can reach some control over its environment, and maximise its chances of success in achieving its goals.

During the 1950s and 60s research on intelligent interaction focused on linguistic-philosophical models of language and communication, while research activities in engineering, AI and computer science centred on computers and computation, developing software and computer architecture as well as investigating the mathematical properties and computational power of algorithms and automata. The main concepts and techniques for the logical manipulation of symbol strings were developed at that time: formal language theory devised properties and well-formedness constraints for different types of symbol strings. It strongly influenced natural language studies, and indeed, the foundations for formal linguistics, and hence computational language processing, was also laid at that time. Chomsky's famous book, *Syntactic Structures*, was published in 1957, and it provided a new methodology for language studies: Chomsky showed that natural language could be described with the help of rules of grammar which guide the algorithmic production of linguistic expressions. The formal and computational properties of natural languages thus became the object of active studies in linguistics, and established foundations for parsing and language technology. At the same time, linguistic research also elaborated the pragmatic and functional analysis of language. Halliday (1961), see also Halliday (1994), emphasised the use of language in its context, and developed Systemic Functional Linguistics that has its starting point in the functional rather than structural language modelling. The organising principle of the theory is the system rather than the structure: language is regarded as a systematic resource for expressing meaning in the context, i.e. a system for meaning potentials, and its description mainly concerns the choices that the language users can make in order to realise a particular linguistic meaning in a given context.

Wittgentein's work on the philosophy of language has been influential in the language and interaction research too. Wittgenstein investigated how language is related to the world, and he contributed to the approach called ordinary language philosophy, which regards the limits of the language also as the bounds of the world which we can know about. In his later work, he introduced a new view of language: that of the language games (Wittgenstein, 1953). The game metaphor emphasises the importance of studying language in use: the meaning of a word depends on the context in which the word is used, and is reflected in the rules that govern the use of ordinary language in actual life. The rules are neither right nor wrong, true or false, but they are useful for the particular applications in which the speakers apply them.

Sharing Wittgenstein's view of the analysis of ordinary language that meaning is revealed through the study of the use of language, Austin further emphasised that language is not only used to inform the partner about true or false states of affairs in the world, but also to change the world towards one's desired state. Austin (1962) developed the theory of speech acts: by uttering a sentence, we perform simultaneous acts of locution, illocution and perlocution, i.e. the act of saying, the act of asserting a meaning and the act of accomplishing a task. The notion of speech act has greatly inspired dialogue modelling, see Section 2.2.2.

The first interactive system that could imitate human conversations was the Eliza system (Weizenbaum, 1966). It is probably one of the most famous conversational computer programs and also provides a basis for present-day chat-bots. The original system plays the role of a psychotherapist and converses with the users (the patient) by inviting them to tell it more about their problems. The system can fool the user for some time by asking relevant sounding questions that give an impression of a nice sympathetic conversational partner. However, Eliza has no real understanding of the utterances or conversational principles: it is based on pattern matching techniques and random selection of a suitable answer depending on the keywords found in the user's utterances. It has no memory of the interaction, either, and ends up repeating itself or producing incoherent answers. The user soon realises that the dialogue leads nowhere: the questions are of the same type, and sometimes they contradict with what the user has just said or had said during the "conversation". What are missing are the two typical aspects of communication: the speakers' intentions, and the general coherence of the dialogue.

Limitations in the techniques and representations and the sheer complexity of the task of building intelligent systems caused progress in AI and computational modelling to be slower than expected. In 1966, the infamous ALPAC report cut funding for automatic language processing in the US, and there came a break of almost a decade before anything significant happened in the field of intelligent interaction.

This was also the time when Kubrick and Clarke created their famous film *2001 A Space Odyssey* (1968). Although it provided fairly accurate predictions of the future state of technology (see e.g. Stork, 1998), the least realistic predictions deal with HAL's communication and reasoning capabilities. To recognize and produce spoken language, emotions, etc. with the same fluency and accuracy as HAL did in the film are still, after 40 years, some of the most difficult challenges for intelligent computer agents.

On the other hand, the earlier unrealistic aims were concretised in the 70s, as practical work advanced many topics which are important and relevant for intelligent interaction. Such topics as frames (Minsky, 1974), symbolic representation and manipulation (Newell & Simon, 1976), scripts (Schank & Abelson, 1977), planning (STRIPS language, Fikes & Nilsson, 1971), dialogue planning (Grosz, 1977), paved the way for a new revival of AI research and natural language interaction.

Expert systems such as LUNAR (Woods et al., 1972) demonstrated intelligent inference in answering questions about moon rock samples and using a large number of rules was able to perform almost as well as the expert on the limited field.

Besides LUNAR, systems such as PARRY (Colby, 1971) and SHRDLU (Winograd, 1972) demonstrated feasibility of natural language-based interaction. PARRY was adapted from Eliza, but exploited an opposite dialogue strategy:

playing the role of a paranoid, it stated its beliefs, fears, and anxieties and thus engaged the user in a conversation as an active listener and questioning party (in fact, the two programs were connected so as to talk to each other and the dialogues appeared quite natural). However, although PARRY seemed to control the dialogue more than Eliza, both of them were built to amuse the users rather than to advance research on natural language processing. On the contrary, SHRDLU is considered to be proof of the concept that it is possible to build a system that understands natural language. The program interpreted natural language commands with respect to a small block world: it could disambiguate references and ask questions of the user, and showed its understanding by manipulating coloured blocks in a block world. In philosophical semantics, Grice (1957) had developed a theory of meaning, and provided an account of how the intended meaning was transferred from the speaker to the listener. Combined with the speech act theory (Austin 1962, Searle, 1979), communication can be regarded as intential action: by uttering a meaningful string of words with the intention of conveying a particular meaning to the partner, the participants are engaged in interaction where the goal is to affect the partner's mental state.

One of the crucial milestones was the insight that formal planning techniques could be applied to speech acts (Cohen and Perrault 1979, Allen and Perrault 1980). Since speech acts can be regarded as a special type of acts, similar planning techniques to those used to infer suitable physical actions can be used in planning communicative acts to reach communicative goals. Dialogues are thus characterized as a sequence of speech acts, and the speakers decide on their next action on the basis of the communicative plan that they have formed for the goal they intend to achieve.

As proof of the concept of dialogue systems, Power (1979) showed that simulated interactions were possible using planning techniques. In the computer-computer simulations, a dialogue game, with moves rather than speech acts, was intiated between two software agents, and the game continued until the participants believed the goal had been achieved, or that it should be abandoned.

At the same time also, speech recognition developed from single word recognition to utterance recognition (Furui, forthcoming), and systems such as HEARSAY-II (Erman et al., 1980) demonstrated speech understanding using several knowledge sources. This kind of work anticipated the building of dialogue systems which could engage themselves in natural language dialogues and plan their interactive behaviour according to general rules.

2.1.2 Experimental Prototypes

As the knowledge of computational techniques and representations, dialogue, language and speech processing increased, it became possible to undertake the development of intelligent and interactive "Thinking Machines" more realistically.

Despite the philosophical controversy about the Thinking Machine, research activities had shown that natural language dialogues between a human and a computer could take place. The research gained momentum and resulted in several projects and specific research frameworks both in Europe and the US. The 1980s was the time for rapid development, and various experimental systems and research prototypes were constructed. The foundations of dialogue modelling were laid, and such concepts as dialogue structure, dialogue act, sub-dialogue, initiative, feedback, goal, intention, mutual belief, focus space, coherence and cooperation were introduced.

The modelling of human understanding and interaction within limited domains paved the way for present day technologies. For instance, one of the first European projects to focus on a natural language dialogue system, the HAMANS-project (Wahlster et al., 1983), took various cooperative principles into account in order to produce the adequate level of responses. In the US, the EES-project (Moore & Swartout, 1992) studied the generation of explanations, including justifications of system actions and clarifications of the user's follow-up questions; and the MINDS-project (Young et al., 1989) focused on the use of pragmatic knowledge sources in order to generate dynamically the expectations of what the user is likely to say next. Both projects produced knowledge-based systems that were designed to model expert interaction. Such expert system projects like MYCIN (Buchanan & Shortliffe, 1984) diagnosing bacterial infections of the blood, also demonstrated the use of the natural language interaction. One of the first multimodal systems was also built at that time: the Put-That-There -system (Bolt, 1980) allowed users to interact with the world through projection on the wall by using speech and pointing gestures. In Japan, the Ministry of International Trade and Industry announced the Fifth Generation Computer Systems programme in 1982, to develop massively parallel computers, which also supported research on natural language engineering.

The dialogue research emphasised the deep understanding of language and communication. An important concept was cooperation: the system should provide the user with helpful information and not only the factual answer, especially if the original request could not be performed. Various aspects of discourse processing were investigated: dialogue structure, coherence, and discourse relations (Grosz & Sidner, 1986; Hobbs, 1979; Reichman, 1985; Hovy, 1988), cooperative answers (Joshi et al., 1984), and misconceptions (McCoy, 1988). The generation of dialogue contributions was also studied (Appelt, 1985), although much of the generation research focused on coherent text-generation and linguistic realisation of messages (see e.g. Zock & Sabah, 1988). From early on, the research also focused on user modelling so as to help the system provide information that would be appropriate to the user's level of expertise (Paris, 1988; Chin, 1989). On the other hand, research on human-human conversation also made great steps during the 1970s and 80s. Clark and colleagues (Clark & Wilkes-Gibbs, 1986; Clark & Schaeffer, 1989) provided a theoretical model of how speakers establish common ground through their contributions and collaborate in utterance production which has influenced dialogue system research, while the approach known as Conversation Analysis studied naturally occurring dialogues and made observations about people's everyday dialogue behaviour (see e.g. Heritage, 1984, 1989).

In AI research, planning techniques (e.g. forward vs. backward chaining) and formalisms such as STRIPS, had been developed in relation to intelligent BDI (Belief-Desire-Intention) agents. BDI agents were designed for knowledge modelling and intention management, and applied to model planning and action sequences in expert systems (Bratman et al., 1988). This led to extensive research into AI-based dialogue management, covering various aspects of rational action, cooperation, planning, intention, teamwork and mutual knowledge, see overviews in Cohen, Morgan & Pollack (1990), Ballim and Wilks (1991).

2.1.3 Large-scale Dialogue Projects

The next decade, the 1990s was the time for large dialogue projects. The results of research on dialogue systems and dialogue models accumulated, dialogue technology advanced together with the technical level of computers, and funding frameworks facilitated the formation of large projects which aimed at combining research with implemented dialogue systems. A shift in aim was also apparent: rather than just implementing a small research prototype that could highlight a particular research idea or demonstrate a new technique, the aim was to build real working systems that could work in a certain restricted domain such as providing information on train or flight timetables, sightseeing and weather, or making hotel and flight bookings. System architectures and dialogue modelling techniques were also further developed, and, e.g. statistical methods became more popular towards the end of the millennium. Coverage of various dialogue phenomena was also widened, especially with respect to spoken communication.

Research was organised within large international and national projects where different dialogue applications were built. Technological aspects also required cooperation among research groups and large international research projects were formed within which prototypes and working applications were built. For example, the EU projects SUNDIAL (McGlashan et al., 1992) and PLUS (Black et al., 1993) exploited communicative principles, pragmatic reasoning and planning techniques as dialogue management techniques, with applications dealing with Yellow Pages information and hotel reservations. Another European project ARISE (Aust et al., 1995) focused on a train timetable system for Dutch, French and Italian, and studied different dialogue strategies and technologies. The French Telecom project produced ARTIMIS, an integrated spoken dialogue system which is based on strict logical reasoning about the participants' beliefs and intentions (Sadek et al., 1997), while the German national project *Verbmobil* (Wahlster et al., 2000) integrated various natural language processing techniques and allowed users to agree on meeting dates and times using natural spoken language. At KTH in Sweden, the Waxholm system was experimental spoken dialogue system which gave information about boat traffic in the Stockholm archipelago (Carlson, 1996). Somewhat later, the TRINDI-project (Larson et al., 2001) established the information update framework in dialogue systems (Larsson & Traum, 2000) and produced the TrindiKit toolkit for experimenting with dialogue systems within the framework, (Larsson et al., 2000).

In Japan, research activities at universities and research laboratories focused on building spoken dialogue systems which would produce natural sounding dialogues with users. For instance, the NTT research laboratory developed the WIT toolkit for real-time dialogue responses and demonstrated Noddy, a talking head which provides feedback to the speaker by nodding at conversationally adequate points (Hirasawa et al., 1999 Nakano et al., 2000). Research activities at ATR, the international research laboratory jointly funded by Japanese companies and the government, supported several different aspects of spoken dialogue technology, from speech processing to dialogue management and automatic language translation. The work culminated in MATRIX, a speech-to-speech translation system that exemplified how multilingual hotel reservations could be made via a spoken language interface (Takezawa et al., 1998). The system incorporated dialogue research especially in utterance disambiguation, and was part of the CSTAR consortium, twinning e.g. with the Verbmobil and the CMU JANUS (Suhm et al., 1995).

The US national research initiative DARPA Communicator supported speech-based, multimodal dialogue system research on a large scale. It assisted in building the basis for interaction technology: system architectures, dialogue management components and system evaluation. The framework produced the Communicator architecture (Galaxy Communicator, Seneff et al., 1998; CMU Communicator, Rudnicky et al., 1999) besides a large number of research papers and reports (see below in subsections). Another agent-based architecture, the Open Agent Architecture (OAA, Martin et al., 1998) was developed at SRI at the same time. This is a flexible distributed computer architecture which has been used, e.g. in the CommandTalk spoken dialogue interface to a battlefield simulator (Stent et al., 1999) and in the Witas project (Lemon et al., 2001), a multimodal dialogue system for instructing and monitoring mobile robots. The TRAINS and its continuation TRIPS project (Allen et al., 1885, 2000) focused on reasoning and planning techniques, and developed an agent-based architecture to handle cooperative dialogue management in a train scheduling domain.

2.1.4   From Written to Spoken Dialogues

While dialogue systems at the beginning of the 1990s used written language, advances in speech technology made speech the predominant input/output modality for dialogue systems in the latter part of the decade (see Furui, forthcoming, for an overview of developments in speech recognition technology). Speech is a natural means of human-human communication, and it is thus legitimate to assume that dialogue systems that can recognise speech support intuitive and easy interaction better than written language or those with menu interface. Although speech may not be the optimal mode of interaction in tasks which require object-manipulation (text editing) or privacy (speaking in public places), development of conversational interfaces has become more reasonable.

Apart from the technological challenges in integrating speech and dialogue technology, spoken language brought in new phenomena to work on: various spontaneous speech phenomena as well as dialogue management strategies to deal with speech recognition errors. Statistical and probabilistic modelling techniques were also widely introduced into spoken dialogue research through the speech community, where these are used as the standard techniques.

The first studies on spoken dialogues dealt with discourse structure and using prosodic cues for discourse and topic segmentation (Grosz & Hirschberg, 1992; Nakatani et al., 1995; Hirschberg & Nakatani, 1996, 1998) and for cue-phrase disambiguation (Hirschberg & Litman, 1993). Soon the research included various conversational speech phenomena: error handling and corrections (Nakatani & Hirscherg, 1993; Stolcke & Shriberg, 1996; Heeman & Allen, 1997; Krahmer et al., 1999; Swerts et al., 2000), intonational characteristics of topic and focus (Swerts & Geluykens, 1994; Terken & Hirschberg, 1994; Steedman, 2000; Krahmer & Swerts, 2006), the role of syntax and prosody in turn-taking (Koiso et al., 1998), incremental understanding of spoken utterances (Nakano et al., 1999) and non-verbal feedback (Hirasawa et al., 1999; Ward & Tsukahara, 2000).

Dialogue modelling also shifted from the hand-crafted rules which describe deep semantic manipulation of natural language input, to shallow processing and experimentation with various machine-learning and statistical techniques. As dialogue management was abstracted on the level of dialogue acts, much research was conducted on the segmentation, recognition and prediction of dialogue acts. The early work was based on written transcripts with probability estimations of the words and word sequences, usually conditioning the probabilities on previous dialogue act sequences (Nagata & Morimoto, 1994; Reithinger & Maier, 1995; Kita et al., 1996), and later also prosodic information of the speech signal (pitch, duration, energy, etc.) was used to recognise the speaker's intentions (Jurafsky et al., 1998; Levin et al., 2000). In general, a wide variation of statistical and machine-learning techniques has been applied to the dialogue act recognition task, ranging from Transformation-Based Learning (Samuel et al., 1998), Neural Networks (Kipp, 1998; Ries, 1999), LVQ (Jokinen et al., 2001), Bayesian models (Keizer et al., 2002) to rule-induction and memory-based algorithm (Lendevai et al., 2003).

In more industrial settings, n-grams were used to identify user requests for call-routing purposes (Chu-Carroll & Carpenter, 1999), while reinforcement learning was introduced to dialogue management (Levin & Pieraccini, 1997, Roy, Pineau & Thurn, 2000) and to learn optimal dialogue strategies (Walker et al., 1998; Litman et al., 2000; Singh et al., 2000), cf. also experiments with Chatbots

(Kearns et al., 2002). Paek & Horvitz (2000) modelled conversations as decision making processes with belief networks representing dialogue uncertainties, and used utility-maximization to select the action with the maximum immediate utility. This optimization method is opposite to the one exploited in the contemporary reinforcement learning where the action that maximizes the sum of utilities over time is taken. In research concerning errors and error management, Shriberg et al. (2005) proposed that the traditional confidence scores may also be replaced by models of the speakers' prosodic behaviour which can provide more accurately predictions of recognition errors than speech recogniser performance scores.

This kind of research led to the current probabilistic dialogue modelling framework, where reinforcement learning is not only used to optimise dialogue management strategies, but also to provide technologies for the actual dialogue management and to the evaluation of different user models by dialogue simulation (Young, 1999; Scheffler & Young, 2000; Roy et al., 2000; Geogila et al., 2005, Rieser & Lemon 2006; Schatzmann et al., 2005, Williams et al., 2008).

2.1.5 Dialogue Corpora

Traditionally dialogue data had been collected using the Wizard-of-Technique (Fraser & Gilbert, 1989), where the users interact with a computer which in fact is played by another human. However, it has been debated whether realistic data can be collected in this way, and how representative the data is concerning different natural language dialogue phenomena. As the collected corpora had often been quite small, and collected for a particular project and for a particular task, it is also been difficult to make reliable generalisations with regard to phenomena that represent the frequent and common properties of the interaction situations, so as to build robust and flexible systems. As the methodologies also shifted from rule-based to statistical and probabilistic modelling techniques, sufficient training and test data were necessary in order to experiment with these techniques.

In order to address these questions and requirements, large projects were established in the 1990's to work on collecting, analysing and annotating dialogue corpora. The special distribution agencies ELDA/ELRA in Europe and LDC in the US were established to facilitate the distribution of the annotated corpora. Some of the widely used spoken dialogue corpora are the Switchboard corpus (Stolcke et al., 2000), collected in the US, which contains casual spoken conversations, and the ICSI meeting corpus which contains recordings in various multiparty meeting situations (Janin et al., 2003). In Europe, the Verbmobil corpus (Jekat et al., 1995) consists of dialogues that deal with scheduling of meetings, while the Edinburgh MapTask corpus (Anderson et al., 1991) was collected originally for studies on certain pronunciation features, but the setup also allows for studies of natural dialogue phenomena. The participants were engaged in an instruction giving task where the "instructor" had to guide the "follower" via a particular path on the map. The participants had slightly mismatching maps, so they had to clarify, discuss and negotiate about the missing or different landmarks. In Japan, similar corpus collection was coordinated by the University of Chiba, resulting in the Japanese MapTask corpus (Ichikawa et al., 1998). The more recent corpus collections have resulted in the multimodal meeting corpora (the AMI-corpus, http://corpus.amiproject.org/) and spontaneous dialogue corpora in Japan (Douglas et al., 2003; Campbell, 2007).

Because of the different underlying tasks and annotation purposes several different dialogue act taxonomies are used in the analysis of the corpora. For instance, the Maptask corpus uses dialogue game and move coding (Carletta et al., 1996), while the DAMSL (Dialogue Act Markup in Several Layers) dialogue act classification (Core & Allen, 1997) is used in Switchboard (Jurafsky et al., 1997) and corpora collected for practical systems exploit their own classification designed for the application in hand (Jekat et al., 1995). Much effort was thus also put into intercoder reliability assessments (Carletta, 1996; DiEugenio & Glass, 2004), and defining dialogue act taxonomies such as DAMSL (Core & Allen, 1997) within the DRI initiative (Carletta et al., 1997). The DRI initiative worked on multilevel annotations and aimed at providing a general definition of dialogue acts. DRI annotations were in accordance with the ideas from Clark and Schaefer (1989) and from Allwood (1976), and emphasised the two-way functionality of the utterances: they have backward and forward looking functions, and they are related to the previous utterances as a reaction to what has been presented. The main discussion points in dialogue act classification and taxonomies are summarised in Traum (2000). Work has also been done on annotation tools, e.g. the MATE tools (Isard et al., 1998), the AnnotationGraph (Bird & Liberman 1999), and Anvil for multimodal analysis (Kipp, 2001). Recently new initiatives for infrastructures, standardisation and interoperativity continue the work on shared resources.

2.1.6 Dialogue Technology

Towards the end of the millennium, dialogue research was mature enough to be called technology and be challenged by the commercial deployment of the systems. However, the starting point for commercial deployment of interactive systems is different from academic research: the goal is to enable automatic interaction between a service and an end user, rather than to model interaction as such. The development of speech-based interactive systems has followed the HMIHY (How May I Help You) -technology (Gorin, Riccardi & Wright, 2007), which relies on statistical classifiers rather than deep language understanding components. Using a large speech corpus, the classifiers are trained to classify the user's utterances into one of the predefined utterance types, while the other system components support accurate speech recognition, recovery strategies from the situations where the input recognition or interpretation have failed, and unambiguous system prompt generation. For an overview of the dialogue industry see Pieraccini (forthcoming).

The application of experimental research to real usage situations thus altered the research challenges from the modelling of complex dialogue phenomena to the practical handling of the shortcomings of the current technology. The commercial deployment of speech-based systems also supports two new aspects in dialogue modelling: usability and performance evaluation of the system, and standardisation of technology and dialogue design. Especially on the industrial side, standards and tools are important and communities such as X+V and SALT have been established in order to develop particular representation languages and tools.

On the other hand, dialogue research has been extended to cover more communicative features, especially non-verbal communication which indicates the agent's cognitive state, emotions, and attitudes (Feldman & Rim, 1991). The topics concern emotional speech processing, facial expressions, body gesture and movement, etc., which in human-computer interaction are often generally called as affective computing. This approach tries to increase the quality of human to computer communication by assigning human-like capabilities, the affect features, to computers (see an overview in Tao & Tan, 2008).

Usability and performance have always been an important part of a dialogue system's design cycle, but the need to compare and assess different technologies brought usability issues forward. Moreover, the commercial viewpoint emphasises that the system should not only function in the required manner, but also provide benefits to the service provider and the end users. For instance, the EAGLES evaluation group surveyed different areas in language technology (Cole et al., 1996), emphasising that evaluation is needed for forming and refining system design, and assessing the impact, usability and effectiveness of overall system performance. User evaluation was also needed in order to obtain feedback on the system from the users, and to compare different systems and provide information for the potential customers on the available systems.

Various evaluations of dialogue system performance were reported (Walker, 1989; Price et al., 1992; Simpson & Fraser,. 1993; Danieli & Gerbino, 1995; Fraser, 1997; Smith, 1998), and the PARADISE work (Walker et al., 1997b) introduced a general framework for evaluating and comparing task-based dialogue systems. The overall objective is to maximize user satisfaction by maximising task success and minimising dialogue cost.

To confirm with totally objective evaluation criteria, and also to reduce the manual work in setting up user evaluations, work has also been conducted with automatic evaluation methods. For instance, the Dialeague evaluation framework in Japan (Hasida et al., 1995) had attempted to simulate dialogue behaviour on a Map Task type dialogue situation where the dialogue systems should "converse" with each other in order to find a shared path through a Tokyo subway map. Later versions of the Dialeague evaluations have taken the user into account and the evaluations are conducted through the web.

Simulated dialogue evaluations have also been used by López-Cózar et al. (2003), as an alternative for collecting real user data which requires a great deal of manual preparation. On the other hand, automatic evaluation only assesses the system's functioning with respect to predefined system requirements.

The emerging dialogue technology also required system evaluation in terms of usability and user satisfaction. A common method for evaluating usability is to interview users in order to ascertain their subjective view of the usability of the system.

This is addressed in HCI studies, where the notion of usability refers to the systems being easy to use, easy to learn and easy to talk to (Wallace & Anderson, 1993). Much of the evaluation research concerns user interfaces without emphasis on dialogue capabilities: e.g. heuristic evaluation (Nielsen, 1994) involves inspection of the interface by special evaluators with respect to recognised usability principles (heuristics) such as visibility of system status, match between the system and the real world, user control, consistency, error prevention, recognition rather than recall, flexibility and efficiency of use, aesthetic design and help.

Practical systems should also be measured with respect to the quality of the service they provide as pointed out by Möller (2002). Evaluation should not only deal with the system's performance as perceived by the users, but also with what the users desire or expect from the system, i.e. there is a need to quantify the value of the system for the users. It is not easy to measure the impact of the system on the user, or the user experience as a whole, but as suggested in (Jokinen and Hurtig, 2006), the quality factors for a spoken interactive system deal with the recognition of communicative principles that support natural interaction as well as trust in the system's ability to reliably provide truthful information.

System evaluation is also related to general standardisation process whereby industrial best-practice standards for interface design and manufacturing are defined. Once specific interaction features get standardised, they start to contribute to the user's expectations of what a useful system should be like.

A different view of dialogue evaluation, but essentially related to the user's perception of an intelligent agent, is offered by the Loebner Prize Competition, established 1991 (http://www.loebner.net/Prizef/loebner-prize.html). This is an interactive system evaluation competition, essentially a modified Turing test where human users converse with a computer system via a keyboard and try to establish whether their partner is human or a program. The goal is thus to separate and rank human and machine partners. The system entries were domain dependent until 1997, then domain free, and in 2009, the competition will be conducted with speech systems. The goal is for the journalists to separate and rank human and machine partners.

2.2 Modelling Approaches

The goal of dialogue modelling is to produce generalisations and models for natural language communication. The object of study includes the speaker's goals and intentions (dialogue acts, moves) and their management in planning and interaction processes (initiatives, confirmations, feedback), information flow (topics, new information) and domain reasoning related to the content of the utterances, the context of interaction (task, activity, speaker roles), and general conversational principles (cooperation, rationality).

There are different approaches to dialogue modelling. The two main approaches, which we call top-down and bottom-up approaches, have grown in different research communities with different scientific goals, and have largely been opposite to each other in their general theoretical orientation. Levinson (1983) marks the difference in the two approaches and calls them discourse modelling and conversation analysis. The controversy is concerned with whether or not formalisation is possible for interaction at all, and whether or not formally describable structures are only post-products of interaction that emerge through the actions of the participants without any pre-specified rules. While we acknowledge the methodological differences between the approaches, we will look at the differences from the point of view of building interactive systems. From this point of view, the main difference lies in the role which a predefined dialogue structure is given in the description and processing of dialogues: the starting point can be either top-down rules which specify the kinds of structures that are possible in interactive situations in general, or bottom-up processes which bind dialogue actions locally into larger structures which are thus kind of by-products of the local "well formedness" constraints. In both approaches, however, dialogues are regarded as purposeful behaviour by the participants, and with the basic interactive units being defined via turn-taking and the participants' exchanges of utterances. Top-down dialogue modelling approach has been the main dialogue management method when building experimental and commercial dialogue systems, and we can distinguish two different techniques to describe dialogues: a grammar-based and an intention-based modelling. The bottom-up approach is represented by corpus-based approaches.

2.2.1 Grammar-based Modelling

The grammar-based approach is based on the observation that dialogues have a number of sequencing regularities (questions are followed by answers, proposals by acceptances, etc.), and the assumption that these regularities can be captured by a "dialogue grammar" (Sinclair & Coulthard, 1975). A dialogue grammar defines dialogue units such as moves, exchanges and transactions, as well as their possible combinations. A transaction can, for example, consist of one or more dialogues, which can consist of initiative-response move sequences, optional feedbacks and embedded sub-dialogues which clarify the topic of the main dialogue. The rules can be expressed as a set of rewriting rules, which define structurally well-formed exchanges and transactions, as shown in Table 2.1.

There is a wealth of research within this approach, ranging from theoretical work (Reichman, 1985; Roulet, 1986; Moeschler, 1989; Scha & Polanyi, 1988) to applications like SUNDIAL (Bilange & Magadur, 1992). Structure-based dialogue modelling can produce rather elaborated grammars. For instance, dialogue models

Table 2.1 A simple dialogue grammar. Rewriting rules use the symbols "+" which means "one or more", "*" which means "zero or more", and parentheses which express optional elements. Italicised end nodes correspond to speech acts

Transaction	→	Dialogue+
Dialogue	→	Initiative Dialogue* Response (Feedback)
Initiative	→	*question | request*
Response	→	*answer | acceptance | rejection*
Feedback	→	*acknowledgement*

based on the Geneva school of discourse studies (Roulet, 1986; Moeschler, 1989) consist of four hierarchically organised levels: transactions, exchanges, interventions, and dialogue acts. Dialogue acts are the basic structural elements, consisting of preconditions (informative or conversational goals, mental state conditions and a dialogue situation) and effects (informative or conversational effects, new mental states and a new dialogue situation). Interventions are made up of one or more dialogue acts, while exchanges are made of interventions and/or exchanges. The dialogue structure is made flexible by allowing an exchange not only to be a two-turn but also a three-turn exchange, and by allowing a response move not only to be a single move but a whole exchange as well. Structure assignment and possible dialogue continuations are planned using dialogue rules. In order to capture different aspects of dialogues, four different types of dialogue rules are used: Grammar Rules (making predictions for the next turn according to the dialogue grammar), Dialogue Control Rules (recognizing a failure situation in which to exercise control), Conversational Rules (maintaining a smooth dialogue in terms of coherence and explicitness), and Corrective Rules (bringing about necessary changes in the dialogue history due to the lack of expressive power in the effects of dialogue acts). The rules operate in the different phases of the dialogue generation cycle.

The popularity of dialogue grammars is partly due to their formal properties which allow for efficient implementation using finite-state automata or context-free grammars. Dialogue management becomes a computationally easy and well-defined task, where structure assignment and prediction of the next element can be handled by well-known algorithms that traverse the state network. The script-based dialogue control (see section 2.3.1) corresponds to dialogue grammar style modelling. In particular, the scripting languages define the states and possible transitions in a similar way as a dialogue grammar.

However, dialogue grammars also have shortcomings which make them less suitable for elaborated dialogue modelling. Two problematic issues are the inflexibility of the structure and the multifunctionality of the utterances. The flexibility of a dialogue grammar can be increased by defining very general rules, but the predictive power of the grammar may thus be lost. For instance, the freedom to

Table 2.2   Example dialogue 1

U1:	I'm looking for restaurants.
S1:	In which area?
U2:	What types of restaurants do you list?
S2:	Indian, Mexican, Chinese, Italian, Thai.
U3:	Try Indian in Rusholme.
S3:	Ok, please wait a minute.

allow exchanges (and not only moves) as responses can accord with the analysis of the dialogues, but it is not clear what is in fact outlawed by such a general grammar. The over-generating grammar rules need to be accompanied by other knowledge sources and different layers of constraints (cf. the four different types of rules above). As for multifunctionality, utterances can often adopt two different structural positions. For instance, in the example dialogue in Table 2.2, the structural position of user contribution U3 is multifunctional: it can be analysed as a request for information based on the list of restaurant types given by the system in System2, but at the same time, it also functions as an answer to the system question in System1, giving the requested location.

Finally, from the point of view of an intelligent computer agent, dialogue grammars fall short of possibilities for handling miscommunication or deviations from the predefined dialogue structure. It is not possible to reason why moves in the exchanges are what they are or what purposes they have in the structure, but other means must be evoked for this.

2.2.2   Intention-based Modelling

An alternative to the structural top-down approach is the intention-based approach which views language as action: utterances not only serve to express propositions, but also perform communicative actions. Rooted in the logic-philosophical work on speech acts by Austin (1962) and Searle (1979), intention-based modelling describes communication in terms of beliefs and intentions of the participants. The goal of the speaker's communicative act is to change the listener's mental state and influence her future actions and attitudes according to the intentions of the speaker (cf. Grice, 1957, 1989). The agents are described as being capable of performing purposeful, goal-directed actions (von Wright, 1971), and their rationality appears in their ability to plan their actions in order to achieve goals, each communicative act is also tied to some underlying plan that the speaker intends to follow. Successful communication means that the hearer recognises the illocutionary force of utterances, i.e. the speaker's purpose in using language, adopts the goal temporarily, and plans her own actions so as to assist the speaker to achieve the underlying goal (or, if not planning to cooperate actively in its achievement, at least plans not to prevent the speaker from achieving the goal).

Utterances may contain specific performative verbs that indicate explicitly the speech act performed (promise, baptise, etc.), but usually the act is to be inferred conventionally on the basis of the utterance context. In a dialogue context, the original concept of speech act is extended and modified to cover various dialogue properties in addition to and instead of the original locutionary, illocutionary and perlocutionary force. Dialogue acts (also termed communicative act, conversational act, dialogue move) describe the dialogue state and include contextual information.

There are different variations of the plan-based approach, mainly emphasising different aspects of the communication. For instance, Bunt (1990, 2000, 2005) defines dialogue acts as a combination of a communicative function and a semantic content, and their effect can be seen via their context-changing potential. The communicative function is thus an operation on the context: when a sentence is uttered, not only is its meaning expressed, but the set of background assumptions is changed as well (Dynamic Interpretation Theory). Bunt also provides a multidimensional taxonomy of dialogue acts with the main functions being task-oriented and control acts. In Bunt and Girard (2005) dialogue act taxonomy is defined, with the idea of having general acts specified as is necessary and possible for the application. In case of ambiguous act, the higher level act in the hierarchy can be assigned so as to still allow dialogue act classification but avoiding the unnecessary unambiguity.

Cohen and Levesque (1990a, 1990b) consider specific utterance events in the context of the speaker's and hearer's mental states, and derive the different effects of the acts from general principles of rational agency and cooperative interaction. As they point out, this view actually renders illocutionary act recognition redundant, since intention recognition is reduced to considerations of rationality. However, illocutionary act types, i.e. dialogue act types, can still be used as convenient labels for the speaker's mental state (Appelt 1985), as is the case in many current dialogue systems.

The view that dialogue acts are a special class of the general communicative acts is also advocated by Allwood (1976) who investigates various enablements and basic principles for linguistic communication. According to him, the speakers are rational agents who act intentionally and voluntarily, have motivation for their actions and are competent in choosing and performing their actions. Allwood (1976, 1977) points out the difficulties in determining illocutionary forces, and suggests that features such as the intended and the actual achieved effects (which are not necessarily the same), overt behaviour, and the context in which the communicative action is performed (resulting in the definition of speech-acts having expressive, evocative and evoked functions). Dialogue acts express and evoke the agents' beliefs and intentions, and also give rise to expectations concerning the course of next actions in communication. The agents' dialogue decisions are constrained by certain normative obligations that underlie successful communication and the agents' coordinated actions. Examples of the application of the principles

Table 2.3   Planning operator for the speech act REQUEST,
instantiated with the requested action "open the door". The
logical operators *want, know* and *cando* describe the speaker's
intention, belief and ability, respectively, *not* is a negation
operator and *open* and *door* predicates that describes entities
and the state of the affairs in the world

Action:	REQUEST(speaker, hearer, open(door))
Preconditions:	want(speaker, open(door))
	not open(door)
	not cando(speaker, open(door))
	know(speaker, cando (hearer, open(door)))
Postconditions:	open(door)

in dialogue situations can be found in Allwood, Traum and Jokinen (2000).
Chapter 3 provides a more detailed discussion of the approach which functions
as the basis and inspiration for the Constructive Dialogue Model in this book.

Since linguistic communication, broken down into dialogue acts, is intentional
action which changes the world, it can be described with the help of dialogue planning operators as the planning and chaining of dialogue acts in the similar fashion
as physical actions (Cohen and Perrault, 1979; Allen and Perrault, 1980). As
shown in Table 2.3, a dialogue act can be described as a planning operator which
defines the preconditions that need to be met in order for the operator to apply, as
well as postconditions which describe what the world will be like after the application of the operator. By common planning techniques, the dialogue act operators
can then be linked into sequences of acts, and dialogue plans can be developed
according to various dialogue strategies in order to achieve the desired goal.

The plan-based approach tends to tie dialogue structure to the task structure,
since the speaker's communicative intentions are embedded in the execution of
her plan which is based on the actions needed to perform the task. However,
although dialogue acts emerge from the actions required by the task, the speakers'
communicative strategies form a separate level of planning, independent from the
domain reasoning. For instance, Litman (1985) separates domain and dialogue
planning levels, and Litman and Allen (1987) specify metaplans, with domain
and discourse plans as parameters, as a way of liberating domain planning from
a strict task structure.

The influential discourse theory of Grosz and Sidner (1986) specifies three
levels of description: linguistic level, intentional level and attentional state.
Each level uses its own organisation which captures different aspects of
discourse management. For instance, the speakers intentions are encoded in the
discourse segment purposes which are linked to each other by dominance and
satisfaction-precedence relations on the intentional level. The attentional state is
modelled by the set of focus spaces that are associated with discourse segments
and which distinguish the most salient objects, properties and relations at a

particular point in the discourse. The levels are independent, yet interconnected, and they work in parallel to determine the well-formedness and coherence of the dialogue. In Sections 3.5 and 3.6 the conversational principles and cooperation are discussed in more detail.

Purely plan-based dialogue management requires complex inferences concerning the speakers' beliefs and plans, and even for simple reactions, reasoning about pre- and postconditions may become too complicated to be computationally feasible. Besides the time and effort needed to design the appropriate reasoning rules, complex reasoning and cooperation may reduce the speed of the system and make the approach less desirable for practical applications which emphasise the efficiency of the system components. However, the TRAINS and TRIPS systems (Allen et al., 1995, 2000) show the feasibility of the approach, and ARTIMIS (Sadek et al., 1997) demonstrates that a commercial application can be built using a purely plan-based rational cooperation approach. The approach has also been applied to collaborative applications like Collagen (Lesch, Rich & Sidner, 1998; Sidner, Boettner & Rich 2000), and virtual multiagent simulations applications (Traum & Rickel, 2002; Traum et al., 2003). Other examples of top-down collaboration and planning, in the framework of logical reasoning and problem-solving, include the Circuit-Fix-It-Shop (Smith and Hipp, 1994) which showed dialogue management in problem-solving situations, and Guinn (1996) who demonstrated mixed-initiative negotiation as a collaborative problem solving strategy between two agents who possess complementary information. The work interleaves communication acts with domain planning: if the agent's knowledge is not sufficient to allow completion of the task, the agent attempts to get the missing information through communication from the partner, and the process continues until the task is successfully resolved. Problem solving itself is described as a proof of statements that represent the world, and the principle of Missing Axiom gives rise to communication: if the agent fails to prove the statements with respect to her knowledge, one way to continue is to try to get the missing information by asking the partner.

2.2.3 Corpus-Based Modelling

Top-down methods have been criticised from both theoretical and methodological perspectives. Their descriptive soundness is challenged because of the obvious diversity of human conversations, while top-down predictions are not considered to be an adequate view of the dialogues at all: the nature of conversation is an interactive process rather than a structural product, and thus the structure is created as the dialogue goes on rather than being "there" for the participants to comply with (Levinson 1983). Levinson further points out that although dialogue acts capture some observable regularities of dialogues and are useful within limited domain automated systems, the basic assumptions about segmenting and sequencing are problematic: this is not the way that humans participate in conversations.

Table 2.4 Example dialogue 2

A0:	Let's meet on Thursday then
B0:	Ok. I'll arrive around 5 pm
A1:	I see. Let's make it 6 pm so you have time to get here
A2:	Fine. I'll be waiting for you in my office
A3:	Well, could we make it 6 pm instead

For instance, dialogue acts are often vague, and in some cases the speaker may intentionally leave the main illocutionary force vague so that the interpretation of what is being said will depend on the partner's reaction. Interpretation of the act will be constructed jointly in the course of the interaction rather than being a selectable property of a single utterance. For example, in dialogue example 2 in Table 2.4, B's response can be informing, promising, or suggesting, but it is possible to infer the appropriate act only when we know A's next reaction. In other words, we need to take into account A's interpretation of what B has said, as well as B's acceptance or correction of A's interpretation of what was just said.

Yet another problem is encountered with the dialogue acts: the segmentation of contributions into single dialogue acts is not always possible. While this is often straightforward, it is also possible that dialogue acts span several turns and speakers. For instance, in dialogue 3 in Table 2.5 the speakers produce the dialogue act jointly by overlapping speech and completing the partner's contributions.

These issues were taken into account in the DRI annotation (Carletta et al., 1997), where backward and forward looking functions link the act to what has been said before and simultaneously also commit the speakers to future actions by producing expectations concerning the possible next acts. Consequently, interpretation of communicative acts is constructed by the participants together in the course of the dialogue, rather than being a property of a single utterance. To complement the top-down planning approaches with observations about naturally occurring dialogues, we discuss corpus-based methods which we simply call the bottom-up approach. As a particular example, we also discuss Conversation Analysis (CA), although it is mostly regarded as a research methodology rather than a dialogue modelling technique. However, it has provide inspiration and insights for experimental studies.

Table 2.5 Example dialogue 3

A:	I'll see you then
B:	yep see you
A:	on Thursday

CA started as a branch of sociology called ethnomethodology whereby the analysis of conversations was a method to study and understand social action and interaction (Garfinkel, 1967). The seminal work by Sacks, Schegloff and Jefferson (1974) on telephone conversations started a new way to see language and language use as a tool in building our social reality. The approach emphasises the study of everyday human-human dialogues which may have no obvious purpose at all. Regularities of language are extracted through observations of how language is used in context and generalisations of conversational phenomena should not be understood as prescriptive causative rules but preferences and expectations of what will follow. The goal of the study is not to provide explicit formal rules that describe dialogue behaviour but rather to provide descriptions of the phenomena that occur in conversations and produce a deeper understanding of how language is used in social interaction and how it constitutes our reality. An overview of the departure points, methodological principles and research results can be found e.g. in Gumperz & Hymes (1972); Goodwin (1981); Heritage (1989); Luff et al. (1990).

Cawsey et al. (1993) remarked that it is unclear how much of the Conversation Analysis type of research can be used in the design of human-computer interaction, although some of its insights provide a useful basis for dialogue research in general. CA has introduced useful terminology for talking about dialogue phenomena including, e.g. adjacency pairs, turn taking, back-channelling, significant pauses, insertion and side sequences ("sub-dialogues"), and opening and closing sequences. The observation that utterances are not continuous but actually contain hesitations, pauses, false starts, and are sometimes left unfinished has also provided new aspects to be modelled. Work on repairs and restarts has been of special significance, likewise the notion of the joint construction of the dialogue structure and conversational meaning as the dialogue structure and conversational meaning. Much interaction research has also dealt with social aspects of communication, and has been influential in revealing how language and its use are linked to the ways we form social groups and create bonds, power relations and institutionalised roles (Gumperz & Hymes, 1972).

It must be emphasised, however, that corpus-based research does not necessarily fall in the CA framework as such. What is shared is the empirical starting point in dialogue data, but the methods and aims are different. While the modelling aims at catching various dependencies among the data elements using formal and computational means, the CA mainly focuses on descriptive generalisations of some interesting conversational phenomena.

As mentioned, there is no obvious way to integrate the CA approach in a dialogue system, i.e. enable dynamic construction of the dialogue and the social reality. It may, of course, be possible that in the future some relevant dialogue behaviour can be simulated on a computer agent through statistical and machine-learning experimentation on signal-level data, and if combined with inter-agent communication, CA-type conversations could also be approximated. Such simulations would then present bottom-up solutions to the issues concerning intelligent computer agents. However, at the moment there are no immediate solutions to this. Of course, data-driven approaches as such are popular in language and dialogue modelling, and have provided some of the widely used component technology (cf. parsers, information extraction, speech recognition and synthesis, HMIHY technology, etc.). It is thus plausible that data-driven approaches could be cascaded or integrated into hybrid models so as to match the complex task of modelling intelligent interaction, cf. the neural models of Miikkulainen (1993) and Wermter, Arevian & Panchev (2000), and the probabilistic approach advocated in Young (2002).

2.2.4 Conversational Principles

Various aspects of rational action also affect communication: the participants act according to general conversational principles such as the maxims of cooperation (Grice, 1975), politeness (Brown & Levinson, 1987), relevance (Sperber & Wilson, 1995), rationality (Cohen & Levesque, 1990b; Gmytrasiewicz & Durfee, 1993), dialogue strategies (Goffman, 1970), etc. For instance, the cooperation principles require that the contributions should be clear, unambiguous, concise, and relevant in the communicative context (Grice, 1975), while the listeners should not unnecessarily hinder the speakers from achieving their goals but plan their own actions so that the speakers can realise their intentions (Allwood, 1976).

Grice's Conversational Maxims (Grice, 1975) have been the standard conversational principles in dialogue system design. For example, Dybkjaer et al. (1996) provide best-practise guidelines for developing practical dialogue systems, and include Gricean maxims in the software design and show how a system designer can refine the system's robustness incrementally according to the principles and the user's behaviour.

Speakers are also bound by social commitments and obligations (Allwood, 1976, 1994), and they need to ground their contributions (Clark & Schaeffer, 1989). Jokinen (1996b, 1996c) provides another view of implementing cooperation in a dialogue system: based on Allwood's Ideal Cooperation and communicative obligations, cooperation is implemented as a series of constraints that restrict possible attitudes to be conveyed to the partner. On the other hand, Traum and Allen (1994) consider various types of obligations which are then implemented as separate action rules.

Politeness (Brown & Levinson, 1987) has usually been part of the dialogue system's conventional "thanks" and "good-bye" acts, but it has been noted that experienced users who just want to get the task done quickly (e.g. ask for the phone number), prefer a blunt and quick system to a polite and slow one, while the situation is the opposite with novice users. Complexity contained in polite behaviour and deeper insights into its functioning as an important part of human social behaviour have been studied more recently, especially in regard to Embodied Conversational Agents and their interaction with human users.

Dialogue strategies are usually modelled in terms of initiatives: which partner introduces a new topic, and has right to control the dialogue (Chu-Carroll, 2000). In dialogue systems, system-initiative, user-initiative, and mixed-initiative interaction strategies have been studied (Chu-Carrol & Brown, 1998; Pieraccini et al., 1997), and the effect of the system's capability to adapt to various strategies (Litman & Pan, 2002).

2.2.5 Collaboration and Teamwork

Rational cooperation has been studied in AI-based BDI-agent interaction (Cohen & Levesque, 1990b), and also in terms of utilitarian cooperation which models the balance between rational but selfish partners who seek for the best strategy to maximize their benefits with least risk taking in the long run (Gmytrasiewicz & Durfee, 1993; Gmytrasiewicz et al., 1995). Horvitz (1987, 1988) introduces the notion of optimising the expected utility of a reasoning system, and also discusses rationality under bounded resources and bounded optimality. Similar principles have also been applied by Hasida et al. (1995) to optimal meaning coding and decoding process between ambiguous language constructions.

Basically, agents have several different desires, some of which they have chosen as their goals. Some goals require other agents' assistance in order to be achieved, and some may require collaboration as part of the task itself (lift a heavy item, dance a waltz, play tennis, etc.). Agents are also assumed to have rather selfish view-points: each agent wants to achieve her own goals first and in the most appropriate way. However, the basic setting of social interaction requires that the agents' selfish view-points fit together: if the agents are to interact a great deal, it is better to cooperate since this pays back best in the long run (Axelrod, 1984). In order to achieve their goals, the agents thus construct a mutual context in which they can interpret the other agents' goals and deliberate on their own plans, and, on the basis of rational reasoning, decide on actions which are intended to provide help to the other agents or collaborate with them.

Collaboration is a type of coordinated activity and requires both collaborative planning and action: the beliefs and intentions of the individual agent are extended to cover the beliefs and intentions that the agent possesses as a member of a group. The three-level model of discourse organisation forms the basis of the SharedPlan approach (Grosz & Sidner, 1990; Grosz & Krautz, 1995) which assumes that the members are committed jointly to bringing about a joint goal within a (partial) SharedPlan, which plan then controls and constrains the agent's future behaviour and individual intentions. SharedPlan is also the basis for Collagen, a collaborative interface agent framework (Lesh et al., 1998). It uses SharedPlan type planning in order to determine the plan that the user is after, then in a collaborative fashion works on a way to help the user to achieve the plan.

Levesque et al. (1990) and Cohen and Levesque (1991) define the joint persistent goal in teamwork analogously to the individual persistent goal, except that the agents have a weak mutual goal to achieve the goal: if they discover privately that the goal is achieved, that it is impossible to achieve or irrelevant, they have to make this mutually known. As pointed out by Grosz and Kraus (1995), the agents cannot opt out from their joint commitment without informing their partner. On the other hand, Galliers (1989) notes that situations where the agents have conflicting goals are common in the real world. In fact, they actually play "a crucial and positive role in the maintenance and evolution of cooperation in social systems". Cooperation is thus not only benevolent in conforming to social norms and obligations, but is seeking actively for the achievement of goals and the resolution of conflict if the agents have conflicting goals. She extends teamwork formalism by the agents' reasoning about obligations.

Finally, it is also useful to distinguish collaboration and cooperation. Cooperation deals with the participants' shared awareness and understanding of what is communicated. It is essential for successful collaboration as it supports the construction of a mutual context in which to exchange information and achieve specific task goals. Allwood's Ideal Cooperation (Allwood, 1976; Allwood et al., 2001) presupposes this type of awareness of the social norms: the partners have a joint goal and they cognitively and ethically consider each other in order to achieve the appropriate reaction in context, and they also trust each other to behave according to similar principles.

2.3 Dialogue Management

After having surveyed different approaches and techniques to dialogue modelling, we now briefly turn to their implementation in interactive systems. Interactive systems range from speech-enabled command interfaces to systems which carry spoken conversations with the user, and the terms *interactive system, dialogue system, communication system* and *speech-based system* are often used interchangeably. We use interactive system as a generic term which refers to any automatic machine that the user can interact with. We restrict interaction to be conducted mainly using natural language, although other, non-verbal aspects of communication such as gestures, facial expressions, and gazing, can also be included as various interaction modes: it is possible to have multimodal interactive systems. The *dialogue system*, however, is a sub-type of interactive systems in that it includes a particular dialogue management component which models the interaction between the user and the system (consequently, it does not follow the design of a reactive system, but includes interaction or communication models, see chapter 1.2.1). These systems can also be called *communicative systems*, especially if they include integration of a variety of knowledge sources and address the challenging issue of new technology and rational interaction. We will mainly use the term "computer agent" to refer to this kind of system, to emphasise the new generation of interactive systems that exemplify the "computer as an agent" -metaphor in practice.

Finally, a *speech-based system* is a specific interactive system which uses speech as the interaction mode; likewise it is possible to talk about gesture-based systems, tactile systems, etc. It is also possible to talk about voice user interfaces (VUI) as opposed to graphical user interfaces (GUI), if the emphasis is on speech vs. mouse-and-menu interface. Analogously, a *spoken dialogue system* is a dialogue system which deals with speech input and output. If there is a need to emphasise the conversational capabilities of the system as opposed to simple commands, the term *(spoken) conversational system* is used.

The Dialogue Manager is a component that manages interaction. It also coordinates the participant's communicative goals, reasoning rules, and communicative principles. It must be emphasised that it creates abstract (not necessarily symbolic) representations from the user input, and the representations are further processed by the other system components. Thus, unlike in human-machine *interfaces*, the input in dialogue systems is not directly related to the output, but is manipulated on different analysis levels which determine the appropriate output depending on the dialogue context.

A typical dialogue system includes components for the following tasks (Figure 2.1):

1. interpretation of the input (Input Analyser);
2. dialogue management (Dialogue Control);
3. response planning (Output Generator)
4. maintenance of dialogue context (Shared Context/Information Storage);
5. task management (Task and application);
6. user modelling (User Models);
7. component interfaces (interfaces and representations).

In Figure 2.1. a general dialogue manager architecture is depicted: the natural language front-end components take care of the interfacing with the user, the

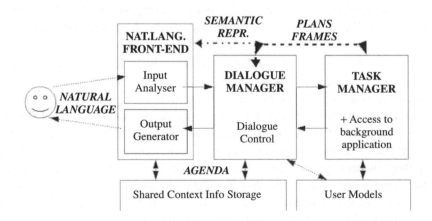

Figure 2.1 A general dialogue management architecture

dialogue manager component(s) handle the reasoning and processes for interaction, and the task manager component(s) concern the application-specific reasoning and processes. The figure also shows special components for user modelling and for storing dialogue history and the shared context built during the interaction.

Some possible data structures for interfacing between the components are also shown: natural language (including verbal and non-verbal signals) between the user and the system, semantic representation that abstracts away from natural language expressions to a more conceptual representation used in the dialogue manager, plans and frames that encode application and domain information needed to reason about the task and what to communicate to the user, and finally agenda that keeps track of the tasks that the dialogue manager needs to execute, including dialogue acts directed to the user as well as internal actions concerning component coordination and updates for the shared context. For a more detailed discussion of the different types of dialogue management architectures see Jokinen & McTear (forthcoming).

2.3.1 Dialogue Control

McTear (2004) describes three different control methods for implementing dialogue systems: scripts, frames, and agents.

The simplest way to build a dialogue system is to use scripts which define possible actions at each dialogue point. They can be implemented as finite state transition networks, where each state corresponds to a possible dialogue state and state transitions describe possible actions to another state. Particular scripting languages such as VoiceXML can be used for rapid development and they also include sub-routines that enable fairly sophisticated dialogue control structures. However, task and dialogue knowledge are only evident through the designer's plan which has been employed in building the system. Each dialogue state and each possible ordering of the dialogue acts must be defined explicitly in the script, and the approach becomes untenable, if more complex dialogues are to be modelled. Moreover, the intertwining of static task knowledge with procedural interaction knowledge makes it difficult to reason with regard to abstract events and relations. It is not possible, except by explicitly enumerating each case, to include inferences or domain reasoning in the interaction management which would be necessary in richer interaction situations.

More flexible dialogue management is possible by separating task knowledge from the possible dialogue actions. This can be done by using forms or frames which define the information needed in order to complete the underlying task. Form-based dialogue management is suitable for tasks where the actions can be executed in various orders and the dialogue can be driven by the information needed at a given dialogue state: the utterances either focus on asking a piece of missing information or provide information to fill the slot. Form-based dialogue management is usually linked to the information state approach to interaction where the dialogue is a way to update the current state of the system so that the requested task (information providing) can be completed (cf. Guinn, 1996, on collaborative problem solving, and Section 5.2.2).

Form-based dialogue management has been implemented, e.g. in systems like GODIS (Larson et al., 2000, 2001), Witas (Lemon et al., 2001), and DIPPER (Bos et al., 2003), which all concern information states and updates.

However, form-based interaction management also has its drawbacks. If the amount of information is large and there are many types of information which are also interdependent on each other, the processing of the form may become a major effort. Also, the underlying task may not be so easily broken down into chunks that can be processed independently, and so the form may require some extra reasoning on how to fill in the slots and what is the best way to receive the missing information. For instance, the RavenClaw (Bohus & Rudnicky, 2003) supports hierarchical task composition, a clear separation between task-specific and more general discourse behaviours, and an expectation agenda which describes what the system can expect at this point.

Finally, the agent-based approach aims at even more flexible interaction management by increasing flexibility with regard to the functioning of the system's components. Based on software agent technology, agent-based architectures support asynchronous and distributed processing. The software agents are independent from each other, and can react to the available information according to their own requirements as defined by the particular task they are designed to do. Agent-based approach also promises easier development of the systems, as well as extension to other domains and tasks, since the new applications can be based on the existing modules and libraries (Hanna et al., 2007). The modular architecture also allows experimenting with the system in a plug-and-play fashion. Moreover, since the software agents, also allow for the use of different techniques, various statistical, machine-learning and rule-based techniques can be used to build hybrid systems.

Examples of various types of agent-based systems include e.g. the GALAXY and CMU Communicator systems (Seneff et al., 1998, 1999; Rudnicky et al., 1999), TRIPS (Blaylock et al., 2002), Interact (Jokinen et al., 2002), and the Queens' Communicator (O'Neill et al., 2004).

2.3.2 Representation

One of the main problems in furnishing computers with intelligent communicative capabilities is the type and representation of knowledge that is needed in order to enable interaction and rational initiative-taking behaviour in the designed artefacts. The argument has centred around whether the system needs explicit representation for its knowledge or whether this can be encoded into the procedures and algorithms that the system executes. For instance, Brooks (1991) argued in favour of a reactive system where a cascaded set of elements communicate with each other but do not exchange any explicit information in any particular format. This view can be related to the grouping of software agents so that their shared activity causes various patterns to emerge in the system's internal state, which configurations then work as an input for the next level agents or the next round of agent operation (cf. section 1.2.3 about implementing the communication model of interaction). A similar view has also been presented in relation to neural networks and subsymbolic processing in general: the result of the combined processing of the individual nodes is more than the simple sum of the processing of each node. In robotics, reactive architecture and physical embodiment of intelligent motor control are some of the active research areas, being developed especially in the context of robots playing a soccer game (see e.g. Behnke et al., 2007; RoboCup, http://www.robocup.org/).

Concerning intelligent interaction, language processing, and cognitive capability, one of the main questions is how the abstract higher level concepts are formed through the clustering and combination of the lower level instances. Interesting research is being conducted concerning how linguistic representations are constituted by sensorimotor activity and experience, although the exact relation between subsymbolic perceptual information and the higher level cognitive processing is not yet clear, see discussion and references in Section 6.1.3.

In machine-learning research the representation of the problem is one of the determining factors for the outcome of the application of the technique (Rendell & Cho, 1990). Discussing different methods for dimensionality reduction concerning document vectors in text classification and clustering, Kaski (1998) points out that the best representation is not dependent on its statistical properties only, but on the empirical discrimination power. Arguments in favour of a certain representation should not thus only concern mathematical properties, but also the ability to produce good results in the classification task. Issues related to different representations and their suitability to various tasks and applications are discussed in the machine-learning papers and textbooks, and e.g. in Witten & Frank, (2005), describing a particular toolkit for machine-learning techniques.

Another important question has dealt with the representation languages. These are data exchange formats which are designed for representing application specific interpretations of user input, and usually they provide standards for representing layered information in a way that is easy to process automatically. A generally used representation language is XML, and its developments and extensions within the W3C consortium (www.w3.org) to such specific areas as semantic information (SISR), speech grammars (SRGS), and multimodality (EMMA) form W3C standards for representing information. Also the speech interface language VoiceXML, mentioned previously, is a representation language for scripting dialogue information, and the agent communication language advocated in FIPA provides a way to present communication between software agents.

However, these languages do not address the issues related to the actual meaning or the semantic content of the elements, but these have to be designed by the system builder separately for the applications. Some research projects tackle knowledge management and knowledge acquisition problems, with the famous example being CYK which aims to equip a computer with human knowledge and reasoning by storing information about encountered facts and their relations explicitly in the computer's memory. On the other hand, Internet is currently the largest repository of electronic data, and its use as the basis for technology that supports interaction and knowledge exchange is widely recognised.

For instance, the Semantic Web (Berners-Lee et al., 2001) aims at knowledge management on a large scale. The data can be organised in hierarchical structures and inheritance frameworks which allow for extraction of the information for the purposes of analysis and automatic processing. Software agents then use the data to perform certain tasks, interacting with each other and with the user. The annotation of the data with knowledge tags is based on conceptual ontologies, defined as a "formal explicit specification of a shared conceptualisation" (Gruber, 1993). One of the interesting ontological questions is related to the dynamically changing world and stability of knowledge. In the context of SemanticWeb updating the knowledge becomes a challenge as the web nodes of the knowledge archives: in the dynamically changing world the web nodes can disappear and new ones appear, while the hierarchy and reasoning which the services and processing of the knowledge are based on, should still be robust and reliable. The problem is not only to add or remove missing links but to re-structure the knowledge, consider e.g. geographical knowledge where the borders of the countries can be changed drastically over the course of the history.

3

Constructive Dialogue Model (CDM)

Human-computer interaction can be seen as a communication situation where the participants have various kinds of intentions and expectations concerning the course and flow of the communication (Allwood, Traum & Jokinen, 2001). In most cases, we cannot, of course, claim that the computer is conscious about its acts and that it understands the meaning of the linguistic symbols it receives and transmits in the same way as the human user: rather, its operation is based on certain algorithms, and its interaction capability is regulated by the rules and tasks which the user wants to perform, such as text editing, information retrieval, planning and programming tasks. However, as argued above, the tasks that the computer is expected to perform have become more complex: the search, manipulation and presentation of information concern problems in situations where the relation between the input and the output is not necessarily predetermined in a straightforward one-to-one manner, but the system appears to reason with no apparent control by the human user. The computer's expressive range has also widened towards sensing and socially aware agents which can recognise and express emotions and affects. The ubiquitous computing paradigm and distributed processing add to the unstructured, uncontrolled nature of interaction as the computer and computing do not form a closed environment any more: it is not with one computer that we interact with but many, and the actual computing may be performed physically somewhere else than in the device next to us. The users also seem to anthropomorphicise the computers and interact with them using the same kind of strategies as with fellow humans. Following the argument that the users assign certain human-like properties to the computers, Nass and Brave (2005) point out that users assign these kinds of human-like qualities to the computer even more if it has a voice: the users react to the tone of voice, the quality, pitch, etc. similarly to how they would react if the speaker were another human. Similar conclusions were also drawn by Jokinen and Hurtig (2006) who evaluated a multimodal system by giving the users slightly different instructions for the same system, differing in whether it was called primarily a speech or a multimodal interface.

Constructive Dialogue Modelling   Kristiina Jokinen
© 2009 John Wiley & Sons, Ltd

In the context of such intelligent computer agent, we aim at studying the enablements and principles of rational and cooperative communication. Communication is regarded as the main factor when characterising intelligent behaviour, and thus it is important that the computer agents also possess adequate communicative capabilities. In this chapter we study the principles of human-human interaction and provide guidelines for intelligent computer agents which would allow natural and flexible communication. The Constructive Dialogue Model is based on the general framework of *Activity-based Communication Analysis* (ACA), which is a pragmatic-philosophical approach to communication developed by Allwood in a series of papers (Allwood, 1976, 1984, 1995, 2000; most recent overview Allwood, 2007). The approach develops ideas concerning human communication as cooperative activity between rational agents, and covers different aspects and levels of communication, applicable to various types of communicative activities.

3.1 Basic Principles of Communication

The purpose of ACA is to provide a framework within which various communicative phenomena can be studied. Expounded by Allwood (1976), the approach has evolved over the course of time, although the basic insights of communication and rational agency have remained the same. The approach has been used in studies of adult language acquisition (Allwood, 1993), linguistic feedback (Allwood, Nivre & Ahlsen, 1992), non-verbal feedback (Cerrato, 2007), cross-cultural communication (Martinovski, 2004), and multimodal annotation (Allwood et al., 2005, 2007). A Situation Theoretic analysis of Allwood's communicative sender activity has been presented by Nivre (1992), and a computational model of ACA, applied to system response planning, has been developed by Jokinen (1995, 1996a, 1996b, 1996c). An overview of the approach and its application to conflict situations can be found in Allwood, Traum and Jokinen (2001).

According to ACA, human communication creates normative social obligations which deal with the participants' evaluation of whether and how they can and want to continue the dialogue, perceive and understand the partner's contribution and react to them in some attitudinally appropriate way. The agents act in accordance with their own goals and they are also capable of attending to the other's goals and abilities. Communication is thus seen as an instrument which serves individual and collective information processing, and it is a part of social activity whose main purpose need not be communicative.

From the point of view of ACA, interactiveness equates to communicative responsiveness which does not only allow, but also obliges the dialogue partners to react to what the other one said. Coherence of the interaction is based on the speakers' need to be understood, and on their obligations to evaluate the partner's utterance and respond in a way that makes the result of the evaluation clear. The reporting of the evaluation must show how the partner's goal and topic have been understood, as well as what is the selected responsive communicative intention in the context.

Formalisation of ACA can go as far as it serves the purpose of building cooperative and natural dialogue systems. It is not necessary to include all the aspects of ACA in the implementation of a practical system: if the goal is to exchange some rather simple and straightforward information, it is usually not necessary to reason about the enablement of communication or the various expressive and evocative attitudes of the sender but rather, code these aspects directly in the interaction management. However, if the task is to model more complicated negotiations, which may include issues of reliabiity and security, the extended communicative capability should be supported by the system.

3.1.1 Levels of Communication

There are several levels of organisation in human communication. ACA distinguishes physical, biological, psychological and social organisational levels, each of which provides necessary, but not sufficient enablement and constraint of communication.

The physical and biological levels of communication concern information exchange on subconscious and automatic levels and they are necessary for enabling communication at a more conscious level and providing neuro-cognitive representations for the agent's observations and sensations. Psychological and social levels concern the observable behaviour of the communicators as individual and intentional agents, and as members of social groups and communities, respectively.

On the physical level, the communicators are regarded as physical entities and their communicative contributions as physical processes where chemical and physical causal laws operate on. Communication deals with human sensory capabilities, such as vision and hearing, and is connected to physical actions that the agent can perform as a result of processing the sensory information. Obviously the development of sensor-based devices will also enable more fine-grained physical communication crucial for intelligent communication with the computing systems.

On the biological level, communicators are seen as biological organisms and their communicative contributions as directed behaviour. Human instinctive reactions to the environment exemplify communication on this level: they are based on an intuitive understanding of the requirements of appropriate actions, without conscious planning and deliberation. Examples in human communication also deal with genetic and physiological messaging as well as with aspects of neuro-cognitive processing studied in biology and neurosciences. From this perspective the level would be better called the neuro-biological communication level. In the light of current research, the distinction between physical and neuro-biological levels need not be so clear-cut when it comes to the actual processing of sensations, but the distinction is clear enough on the general levels of communication enablement, which warrants their separation as different communicative levels.

The levels of communication which are generally acknowledged as the core of human communication are the psychological and social levels of communication. The psychological level of communication focuses on human cognition (perception, understanding, emotion) and on the processes that are associated with communicative capability (motivation, rationality, agency). Communicators are regarded as perceiving, understanding and emotional beings and their communicative contributions are perceptually comprehensible and emotionally charged phenomena. Being motivated, rational agents, they produce communicative contributions which are motivated rational acts.

The social level of communication concerns groups and communities, i.e. aspects of social dynamics that are realised in culture, language, social institutions and activities, shared tasks, coordination and communication. The communicators are members of a society, a particular culture, social institution and linguistic community, and accordingly their communicative contributions are cultural, social, institutional and linguistic acts. They also play certain roles in social activities, and their communicative acts are contributions to that activity through their role.

From the point of view of human-computer interaction, the levels may appear to be too wide and maybe unnecessary since the initial setting so far has been more limited, and the metaphor of the computer as a tool does not lend itself to designing such full-blown communication. Considering communication in the context of smart environments where different tasks are performed automatically with the help of sensing and interactive technology, the systems tend to be regarded as communicating agents and the levels do not appear to be as irrelevant as one might first think. In fact, successful human-centred design of ubiquitous, intelligent applications seems to offer analogies where the enablements and constraints of interaction can be taken into consideration. First, language-based communication is a complex process which includes both symbolic and sub-symbolic organisational principles and presupposes that information exchange functions on different levels of communication. We also note that the development of methods, techniques and applications for intelligent interactive agents can benefit from the analogy to human communication enablement, and the different communicative levels can be transferred into their design including agent networks that deal with intra- and inter-device communication.

For instance, we can say that the physical level of communication involves physical enablement for interaction, such as mouse and keyboard, screen size and its brightness, recognition and synthesis of speech signals, gestures, facial expressions, etc., where the physical properties of sound, light, touch and movement form the sensory basis of interaction. The physical level can also be said to concerns the ergonomics of computer devices and interfaces. Various experiments on typing, navigation, the user's reaction-times, and cognitive load have shown how human perception and cognitive constraints affect successful interaction, while the novel interaction techniques like touch-screens, gesture recognizers, eye-trackers, etc. bring in more options (i.e. modalities), for the physical communication level. Furthermore, we can say that the networked communication infrastructure as well as with mobile and location-based technology does not only enable communication between humans, or between humans and various services and applications but also functions as the physical communication level for intelligent computer agents.

There are no straightforward analogues for the biological level of communication in HCI, since this level of communication mainly concerns the communicator's internal activities and subconscious adaptation to the context. We can, of course, use biologically-inspired computing algorithms, such as Neural Networks, Genetic Algorithms, or Ant Colony Optimization, to model some aspects of neuro-biological communication. However, these algorithms serve basically as efficient ways for handling computationally complex problems rather than as models of the processes on the neuro-biological communication level. On the other hand, adaptive and learning systems that can change their behaviour according to different environmental situations come close to the self-regulating and self-organising systems that communicate on a level analogous to human neuro-biological communication. The main question, then, concerns the difference in the platforms that enable such behaviour: neurons implemented in carbon and silicon differ in their properties, and it is difficult to say if the processes conducted by silicon-based neurons in computers have the same qualitative results as those conducted by carbon systems, i.e. humans. Answers to this question create philosophical disputation that continues the one which the notions of strong and weak artificial intelligence generated in the early years of the research: can machines have consciousness, what does intelligence and agency actually include? We will not go into the details of this disputation, but the interested reader is referred to Kurzweil (1999) or the publications of the Singularity Institute for Artificial Intelligence (http://singinst.org/research/publications).

In the general setting of the organisation of communication, we now proceed to studying the actual communicators and the prerequisites for successful communication. The assumption is that the communicators are rational agents and the driving force in their interaction is the set of intentions related to the underlying task that they want to complete. Communication is enabled by mutual willingness and cooperation.

3.1.2 Rational Agency

Following Allwood (1976), we can talk about linguistic communication as rational and cooperative activity. By introducing the principles that capture the meaning of normal, rational agency, Allwood defines the concept of *Ideal Cooperation*. He proceeds by defining the type of communication that is typical among human beings, *Full-blown Cooperative Communication*, as Ideal Cooperation between Normal Rational Agents. By participating in the communicative activity, participants have certain roles which further determine their communicative activity.

Furthermore, communication creates normative social obligations, in terms of which *Communicative Responsiveness* can be defined.

According to Allwood (1976), rational motivated action can be analysed with respect to "norms that an individual agent tries to follow in his own behaviour". These norms or principles concern the actor's agency (i.e. the actor's intention and purpose for the action, and her voluntary acting), normality (the actor's motivation for the action, and her decision to seek pleasure and avoid pain), and finally rationality (i.e. the actor's adequate and competent behaviour).

Galliers (1989) points out that Allwood (1976) provides one of the few attempts to summarise the traits of a rational agent. Usually agency refers to properties related to intentional and independent deliberation: the agent's actions are purposeful and there is a free choice among the alternatives. Allwood summarises the traits of a rational agent into a principle that says "Typical human beings are normal rational agents", and which functions as a precondition for the other principles. The principles of normality, rationality and agency do not describe the properties of the agents, but rather, they are to be understood as "statements of assumptions that typical socialized agents make about the behaviour of other individuals".

Agency refers to the assumptions that agents should have a reason for their actions, a genuine opportunity to choose between different alternative actions, so that their decision making is free. The agent's behaviour is thus intentional and purposeful and intentionally controlled. Allwood states the principle as: "The intentionally controllable behavior of an agent is intentional and purposeful."

The principle of voluntary action states that "The actions of an agent are not performed against his own will". Since there is a trivial sense in which every action is voluntary, the negative formulation of the principle is used to emphasise the stronger sense of "voluntary", namely: an action is voluntary if and only if an agent does not believe that she would be in danger if she does not perform the action. The stronger sense of "voluntary" allows Allwood to distinguish involuntary action from voluntary, his example of the first one being forced labour at gun-point in a concentration camp.

The normality principle states: "The actions of a normal agent are motivated, and that "Normal agents do not act so as to decrease their pleasure and increase their pain". Motivation concerns the agent's internal need, desire, or want, but excludes the external circumstances that give rise to the motive. The pleasure and pain principle is expressed in a negative form since a positive formulation "agents strive to increase their pleasure and decrease their pain" is considered to be too strong.

Finally, rationality is defined with respect to the agent's competence to select adequate actions. Allwood states the adequacy principle as follows: "The actions of a rational agent are selected so as to provide the most adequate and efficient way of achieving the purpose for which they are intended", and the competence principle: "The actions of a rational agent are performed only if he thinks it is possible to achieve their intended purpose." In other words, the actions of a rational agent are selected so as to provide the most adequate and efficient way of achieving the purpose for which they are intended. Moreover, the actions of a rational agent are performed only if the agent thinks it is possible to achieve their intended purpose. Rationality is thus related to the agent's observations of the environment and other agents and is effectively a sign of the agent's ability to plan, coordinate, and choose actions so that her behavior appears to be competent and realises the goal(s) which motivated the action.

It is tempting to view rationality as a property that agents may or may not possess, and try to find suitable measurements for rationality of their behaviour. However, as emphasised above, we do not claim that agents actually act rationally but only that they act in a way that seems rational to them and to the partner. Rationality emerges as an effect of the agent's successful goal realisation, and it is learnt in interaction with the environment. The acts and act sequences that the agent gets rewarded by successful achievement of her goals are considered "rational" for their purpose and stored for future use. Better strategies, i.e. more rational behaviour patterns can be learnt by observing the other agents' acting, and by exploring and refining one's own learnt patterns. Rationality is thus not a property of a particular act either, but related to the agent's goals and circumstances. It is encoded in the reasoning processes used to achieve the goal: it is a procedural concept that does not exist without the activity that drives interaction and communication in the first place.

However, rationality describes an act's context changing potential, not the actual contextual changes: the actual effects need not be exactly as intended (the answer may be evasive, an attempt to rent a car unsuccessful), although the act itself may be considered rational. The agent must thus learn to distinguish between an abstract plan as a recipe to achieve a desired state, and the execution of the plan in actual circumstances which may cause the plan to fail. Here we enter the area of the agent's resource limitations: the agent is resource-bound and cannot know all the factors that influence the intended effects at the time of acting.

On the other hand, Allwood (1976) also points out that achieving the intended effects does not render an act rational either. For instance, ordering a taxi without the intention to go somewhere is considered irrational (or at least a bad joke) even though the effect of the request is achieved when the taxi arrives. Rationality is thus tied to the act's assumed function in a larger context: rational acts are instruments in achieving some goal.

The question now arises, how much of the rational, normal agency can be assumed to be applicable in human-computer interaction. Especially, if the computer is understood as an agent, how much intentional and purposeful behaviour, motivated action and rationality can be directly applied to human-computer interaction. We spell them out as rules which govern the system's reasoning: determining joint purpose and checking the system's communicative obligations.

3.1.3 Ideal Cooperation

Participants are engaged in Ideal cooperation, if the following constraints are observed:

- joint purpose;
- cognitive consideration;
- ethical consideration;
- trust.

Joint purpose refers to the fact that the participants have common purposes that they strive voluntarily to achieve. The purposes can range from specific task-related goals like "know when the next bus leaves" to general phatic goals that aim at maintaining social relations or keeping the channel open. Sometimes the agents pursue their own private purposes and cooperation is only partial. However, as long as the agents are communicating, they have at least one common purpose: transfer of information. This purpose is the basis of communication, and if it disappears, there will be no communication. This idea also forms the basis for the NewInfo-based response planning which will be discussed further in Chapter 5.2.2: response planning starts from NewInfo which encodes the new information to be communicated to the user and if there is no new information to be exchanged, there is no need to communicate either. It shoud be emphasised that the "transfer of information" is understood widely here, and also includes phatic communication (Malinowski, 1923), the establishment and maintenance of communicative contact. When agents engage themselves in communicative activity, they also create social bonds which can vary depending on the communicative situation and the roles the agents assume for themselves. However, one of the agents' first tasks is to determine and coordinate the joint purpose for the dialogue, and this purpose can include aspects of phatic communication such as the management of social relations, and opening up and keeping the channel open. This is also an important part of the system's reasoning which will be discussed more in Chapters 4 and 5.

The two consideration constraints, cognitive and ethical consideration, encode the core reasoning for the evaluation and planning of communicative acts: the agents are ethically and cognitively considering each other in trying to achieve the joint purposes. Cognitive consideration deals with the agents' perceptual and cognitive activity: agents analyse and evaluate the contributions with respect to their own intentions and what they have observed and know about the partner, the communicative situation, the world in general. Ethical consideration is the main contribution of Allwood's theory to the discussion of rational cooperation. While cognitive consideration guarantees the agent knows about the other agent's goal, ethical consideration requires that the agent treat the partner in a considerate and ethical way. In order to show cooperation it is not enough that the agent is aware of another agent's goal: the agent should not do anything that would prevent the other agent from acting as a rational, motivated agent too.

Cognitive consideration refers to the epistemic side of rationality: the agent's knowledge of what is a rational way to achieve joint purposes. On the basis of this knowledge, the agent can predict the partner's reaction and plan her own acts so as to follow the principles of normal, rational agency. However, Allwood also emphasises the ethical aspect of the acts: the agent should not only act according to some norms, but also not act so that other agents are unable to maintain their rationality. The ethical dimension also provides a counter-force to epistemic rationality which is often insufficient in accounting for the rationality of apparently irrational actions. An agent may choose to increase pain instead of pleasure (e.g. not disclose information which would save her own face but cause harm or embarrasment for somebody else), or choose a method which is both inefficient and inadequate for a particular task (e.g. follow instructions of a superior knowing that they involve a long detour), but still her behaviour can be described as rational since it shows consideration towards the overall social situation. Rationality of such seemingly irrational actions is accounted for by the act's ethics: the agent should not only attempt to achieve her goals in the most rational way, but she should not prevent other agents from realising their goals either. We can say that the agent's acts are considered rational if they successfully fulfill the goal (operational appropriateness) and do not cause obstacles to other agents (ethical acceptability).

The last constraint on Ideal Cooperation is trust: the agents trust each other to act according to the other constraints, unless they give each other explicit notice that they are not. In order for the cooperation to work, the participants assume that the partner acts according to the other three constraints, and if the assumption is not valid, the partner must notify the agent of this. The problem that the agent does not have access to the partner's mental processes or attitudes is thus alleviated by assuming that all communicating agents share the same principles for cooperation. Consequently, if the agent does introspection about her own mental process, this is sufficient to enable her to draw conclusions about the partner's mental processes as well. Of course, the partner may be lying or acting maliciously, but as long as there is no overt signalling of this, communication proceeds as if the constraints of ideal cooperation are fulfilled.

The creation of trust is one of the challenges for human-computer interaction, and especially for spoken dialogue systems that appear in the intelligent environment. As the requirements for interaction design change from the designer's "omniscient" view into the communicating agent's "minimal rationality" view, in which both human and computer agent have partial knowledge of the world, they need to trust each other so as to act according to the same principles of Ideal Cooperation and provide truthful information that would allow the partner to operate in an approriate manner.

As mentioned above, the agents do no necessarily fully cooperate all the time. Moreover, cooperation does not mean that the agents always try to react in the way the partner intends to evoke. As Galliers (1989) points out, if agents are always in agreement and ready to adopt the other's goals, they are benevolent rather than cooperative. Ideal Cooperation captures the agents' basic attitude: their actions are based on willingness to receive, evaluate and react to the partner's contributions. Communication may include pursuing a conflict with the partner, but if such a conflict becomes so serious that it makes any cooperation impossible, communication will break down as well.

We must also note that dialogue collaboration is shown on different levels. It is demonstrated on working together on a particular task, on exchanging information, co-producing utterances, and adapting to the partner's lexical items and speaking style. Computational research talks about grounding and construction of shared context (Clark & Wilkes-Gibbs, 1986; Traum, 1994; Fais, 1984; Jokinen et al., 1998), whereas in psycholinguistic research this kind of adaptation is called alignment (Pickering & Garrod, 2004).

3.1.4   Relativised Nature of Rationality

As pointed out by Cherniak (1986), rational agents are not omnipotent and omniscient agents but act according to minimal rationality conditions. In fact, the "rationality" of the rational agent can be seen in their ability to restrict their reasoning onto a smaller set of pertinent statements of the world, whereas the notion of perfectly rational agent poses a computationally and cognitively intractable problem: there are cognitive limitations to perform arbitrarily large computations in constant time, and it is unlikely that the agent will, and can, check the logical consistency and coherence of all their beliefs whenever a decision needs to be drawn to act. To construct less idealised models of rational agents, Cherniak further explores minimal rationality, i.e. the minimal level of rationality that is executed by the agents in practise so as still to be considered as a rational agent. Resource-boundedness was taken as a starting point for the BDI-agents (Bratman et al., 1988).

Given the finiteness of the agent's cognitive capacity, the problem of choice also arises: which of the all possible inferences (sound and feasible inferences that the agent can produce) are operationally appropriate (focused on in a given situation), and which of the appropriate ones are also ethically acceptable (allow the partner to maintain her rationality) to be undertaken in the given context. Agents can opt for sub-optimal action (action that is not maximally appropriate), if this is believed to be more acceptable in the context, and for unethical action (not maximally acceptable), if this is believed to be more appropriate in the context. The tradeoff between the two extremes is tied to the agents' values and commitments, and can be modelled by a cost-reward mechanism where inferences are assigned costs or rewards depending on whether or not they are based on the beliefs which the agent assumes to be currently focused on and which she has committed herself to. The agent's commitment to the particular belief set which she considers pertinent to the situation is important, since the agent can hold a set of beliefs which is inconsistent as a whole. She can base rationality judgments on any consistent subset of her beliefs, since the consistent set of beliefs is not sufficient for the implied action to be undertaken: it is not contradictory to describe an act as rational, yet not undertake it. For instance, one may understand that the affirmative action clause in job announcements is rational from the point of view of women's under-representation in higher managerial posts, yet not act according to it in practice. The agent can understand the rationality of an act on the epistemic level, but to undertake such an act, she must commit herself to the beliefs that form the basis for reasoning. Conversely, if the agent is committed to the beliefs but acts against the implied conclusion, her action is considered irrational.

On the other hand, an act can also be described as rational to different degrees depending on how well the agent thinks the act fits goal fulfilment and ethics. If an act is inappropriate for the purposes it's intended for, or unacceptable on the basis of ethical consideration, its rationality can be questioned, and if the act is perceived as both inappropriate and unacceptable, it is deemed irrational. Thus the opposite of rationality is not necessarily irrationality, but an increasing non-rationality which changes to irrationality at some non-fixed point where acts can no more be considered either appropriate or acceptable.

Although planning is needed for any successful action, the act's rationality can usually be assessed retrospectively: no careful deliberation can guarantee the success of an act due to incompleteness of the agent's knowledge at a given time. A statement that an act is rational is a value judgment of the act's appropriateness and acceptability rather than an objective description of the state of affairs. It tells us that the observed act is understood on an epistemic level and accepted on an ethical level, but not that the act actually is rational. It may be that the agent lacks information that would render the act irrational, such as when an agent draws a detailed map to help the partner to find a restaurant, but later learns that the restaurant has been closed down and does not exist anymore; the agent's drawing of the map is still considered rational, although it later turned out to be a waste of time. Given the finiteness of cognitive capacity, the agent cannot guarantee the absolute rationality of her actions, only their relative rationality: the agent acts in a way which she believes is rational in a given situation. This is often understood by the agents themselves by adding a relativising comment "according to my best knowledge".

The weaker view of the agent's rationality is also related to the fact that the agent may believe that an act is rational without this being the case for the partner. Rationality is relative to the goals of each agent, but it is also relative to the agent's observations about the tasks and acts that are performed by the other agents. If the agents then start to discuss about the rationality of some particular acts, the discussion will take place on the metalevel and rationality will be subject to dynamic, interactive and intersubjective relativisation. Rationality, being a judgment of the manner in which the agent attempts to reach her goal, is negotiable: the agents can argue about the appropriateness and acceptability of a particular act by requiring and providing explanations, motivations and reasons. Moreover, this metalevel communication can be recursive: the appropriateness and acceptability of the explanations, motivations and reasons can also be questioned. Hence, the agents can affect each other's views and opinions of what is considered rational acting, and consequently, modify or even change their reasoning, depending on whether convincing arguments can be provided that warrant the change.

This ties rationality to a more general context of cultural values and social norms: rationality judgments are often shared by the community members, because the more or less uniform environment tends to favour particular types of actions over others. We will not push this line of argumentation further, although we want to emphasise the role that social aspects play in rationality considerations: although an agent's action may be successful in her goal satisfaction and acceptable with respect to individual people, it need not look rational from the community's viewpoint, and thus "one-sided" rationality may be mutual irrationality, ranging from eccentricity to lunacy.

3.2 Full-blown Communication

Communication is related to the sending and receiving of information from one agent to another. In order for the agents to communicate in the first place, some basic requirements must be met. Allwood (1976) discusses four necessary requirements for communication are distinguished[1]:

- contact;
- perception;
- understanding;
- attitudinal reactions.

Contact indicates that the agent shows or elicits willingness to establish or maintain contact and to go on in the communication, while perception indicates that the agent shows to have perceived or elicits signs of the partner having perceived the message. Understanding indicates that the agent shows to have understood or elicits signs of the partner having understood the message. Attitudinal reactions concern the agent's actual response that expresses the agent's intentions and attitudes, and also elicits further reaction from the partner.

The basic categories of Contact and Perception also correspond to what Clark and Schaefer (1989) have called *acknowledgement*, i.e. the agent signalling to the partner that his contribution has been understood well enough to allow the conversation to proceed. It should also be noticed that Contact need not be face-to-face (as e.g. with TV presenters) and Perception need not be conscious (unintended communication).

---

[1] Allwood calls these four requirements *communicative functions* which is unfortunate because of the ambiguity of the term with respect to speech act theory.

Contact and Perception are preconditions for Understanding and Reactions, which involve higher-level intentional reasoning. Different communicative conditions can thus occur depending on how the three basic requirements are fulfilled: only contact (C), both contact and perception (CP), contact, perception and understaning (CPU), and the full communication with contact, perception, understanding and reaction (CPUR).

Allwood (1976) discusses different types of communication, and defines typical human communication as a basically cooperative activity, where all the requirements are met. This type of communication is called *Full-blown Cooperative Communication*. When talking about "communication" in this book, we refer to "full-blown communication".

Full-blown communication takes place between (human) agents with some degree of conscious awareness when the sender's signal is apprehended as a sign by the receiver.

The dialogue agents also act so that their actions manifest rational thinking and they follow the principles of ideal cooperation. After each contribution they evaluate whether the requirements for communication are still valid: they must consider if they can and want to continue, perceive and understand the contribution, and then produce a reaction that shows cognitive and ethical consideration towards the partner. However, the agents can respond in different ways depending on which level of requirements they want to respond to, or whether they want to respond positively or negatively, explicitly or implicitly. They can give *feedback* on the evaluation on different levels. The feedback on contact and perception is usually referred to as backchannelling, and it is widely studied in spoken language research. Often the CPU-type feedback is given by subtle signals of gesturing, eye-gazing and body movements. References, and discussion in the ACA framework, can be found e.g. in Allwood, Nivre & Ahlsen (1992); Allwood et al. (2007).

Communication is an interactive task. Therefore, each communicative contribution also serves communication management purposes which deal with the general goals of the agent's rational action. Allwood (1994) divides communication management into two kinds: the speaker's own communication management and interactive management. The speaker's own communication management deals with the speaker's on-line planning: correcting and changing one's contribution while communicating. Interactive management concerns turn taking (distribution of the right to speak), sequencing (structuring the dialogue into sequences, sub-activities, topics, etc.) and feedback (elicitation and giving of information about the four basic communicative requirements).

3.2.1 Communicative Responsiveness

Communication creates normative social obligations, which deal with issues such as people's availability and contactability for information coordination and their willingness to report on the results of the evaluation of the information. These obligations can be understood as *obligations of responsiveness*, where responsiveness is itself a consequence of the human ability for rational coordinated interaction (Allwood, 1992).

Communicative responsiveness can be associated with what have been called *preferred responses* (Schegloff & Sacks, 1973) or *preferreds* (Levinson, 1983). This means that a communicative act often puts a certain pressure on the hearer to react in a certain way. For example, an initial greeting gives rise to a pressure to respond with another greeting, and an expression of gratitude to respond with a disclaimer. However, as claimed in (Allwood, 1984), these concepts do not capture the normative aspect of responses which underlies communication in the first place. It is the obligation of responsiveness which requires the hearer to evaluate the utterance and respond to it in an appropriate, socially committed way.

Responsiveness is an essential part of being cooperative: being able and willing to respond is shown in the dialogue participation by appropriate responses. In particular, it is tied to the activities and roles that the dialogue partners are engaged in. For instance, one is under different pressure when answering the questions of one's manager than when answering the questions of a nosy friend. The employee role thus reinforces the responsiveness obligation in relation to the manager.

Levinson (1983) points out that a question usually requires a response, but the response need not be a direct answer. For instance, a question like *What does John do for a living?* can be happily followed by partial answer ("Oh, this and that"), rejection of the presuppositions of the question ("He doesn't"), statement of ignorance ("I've no idea"), or denial of the relevance of the question ("What's that got to do with it?"). While the first three responses can be called preferreds, the last one is a dispreferred response in that it is not expected as it requires the questioner to provide further explanation for her question.

What makes an utterance after a question function as an answer is that it occurs after the question with a particular content. However, as shown above by the variety of response types, the responses are reactions to a dialogue situation rather than just any contribution following a question. In ACA terms, we say that the response is evoked by the obligation of responsiveness, and its functioning as an answer to a question is based on the agent's trust that the partner is a rational agent who replies in a considerate way. The relevance of the response in its context is evaluated by the agent according to her communicative intentions and knowledge about the context and world. For instance, a partial answer may trigger further clarification if the agent intended to know about John's actual employment history, or it may be a satisfactory answer if the agent was not interested in further details in the first place. In both cases, however, the answer is considered a rational reaction on behalf of the partner, and the agent is obliged to evaluate it and provide a reaction on her behalf. (Cf. the presentation-acceptance cycle introduced by Clark & Schaefer, 1989).

Table 3.1 Expressive and evocative dimensions of four communicative acts. From Allwood (1992)

Type of communicative act	Expressive dimension	Evocative dimension
Statement	belief	(that listener shares) the belief judgment
Question	desire for information	(that listener provides) the desired information
Request	desire for X	(that listener provides) X
Exclamation	any attitude	(that listener attends to) the attitude

Besides providing different factual answers, the agents' responses also address the CPU-requirements discussed in the previous section: the agents are obliged to give feedback on the success of their communication.

The obligation of responsiveness also plays a role in the accounts of communicative acts. According to Allwood (1976), each communicative act carries both expressive and evocative dimensions. The *expressive dimension* deals with the expression of an attitude on the part of the speaker, and the *evocative dimension* with the evocation of a reaction in the listener. By the expressive content of a communicative act the speaker provides information about her mental state (emotions and attitudes), and also about her physical and social identity. By the evocative content of a communicative act the speaker intends to influence the hearer's mental state, and at very least wants the hearer to apprehend some information. The effects of an utterance and the intentions behind them are also distinguished. This distinction corresponds to that between the actually evoked response and the evocative intention of an utterance, respectively.

Allwood points out that these functions must not be confused with the illocutionary and perlocutionary forces of speech acts. Rather, Austin's concept of illocutionary force is split up into three dimensions of expression, evocation and obligation of responsiveness, and his notion of perlocution corresponds to what is actually achieved by a communicative act (evoked response).

Table 3.1, taken from Allwood (1992), summarises the stereotypical expressive and evocative functions of the four conventional communicative acts.

3.2.2 Roles and Activities of Participants

By virtue of participating in activities, the participants occupy or play roles which are constituted by the activities. The roles can be characterised by global communicative rights and obligations that the participants have in the different dialogue situations and the participants can have several roles and their roles can also change. The information giver has different obligations from the information receiver, and the roles are further differentiated when, e.g. talking to a stranger or to one's friend.

From the user's point of view, interaction with a computer agent is a social activity in which the user is engaged: the user may need to obtain information or some services, or may be talking to a robot pet. The user is typically involved in two related activities: communicative activity with the computer agent and non-communicative activity with respect to the task which the dialogue serves. The activities are related, but in some cases more task-related planning is necessary (e.g. problem solving situations), and in some cases focus is more on the communicative activity (e.g. building social relations).

In the present-day interactive systems the two communicative activities are usually combined, and the dialogues follow task completion steps which the user is assumed to be familiar with. However, for more flexible interactions, it is important to address requirements imposed by both communicative and non-communicative activities, i.e. provide factual task-related information in a manner that conforms to cooperative communication. From the point of view of ACA, this requires analysis of the activities in which the user can be engaged with the computer agent, and specifying typical roles in these activities so as to provide accounts of the participants communicative rights and obligations.

The roles that the user assumes for herself and for the computer can usually be characterised as a master and a servant (or a user and an inanimate tool): the user initiates a task (e.g. asks for tourist information or wants to make a reservation), and the system tries to fulfil the request. In principle, the user has no communicative obligations when engaged in a dialogue with a computer agent either: she can choose not to initiate a dialogue or to terminate a dialogue at any point. However, if she plays the role of an earnest dialogue participant, the dialogue constitutes a situation where she is constrained by local communicative obligations that are related to her being a cooperative rational agent. These obligations include willingness to provide the information (specification, clarification) requested by the system in order to complete the task that the dialogue was initiated for and willingness to terminate the dialogue explicitly. The obligation of explicit closure of a dialogue is supported by the need for explicit communication management, i.e. by the need to plan and coordinate one's tasks and intentions with other agents. On the other hand, the user has the right to continue interaction as long as she likes – and is willing to pay (mobile phone bills).

The computer agent's role in practical applications is to be a helpful assistant or a reliable friend who provides truthful information, but not necessarily take initiative or make decision on behalf of the user. On a general level, this means that the computer agent always ready to accept user contributions as input, produce their analysis and provide the user with appropriate responses, but its communicative activity may be restricted with respect to introducing new topics. On the other hand, it should be pointed out that the agent's role as well as communicative rights and obligations are important when considering possible manifestations of the agent's intelligence: it is difficult to demonstrate intelligent behaviour in simple information providing tasks playing roles with limited communicative freedom. An important question thus deals with what kind of communicative rights and obligations can, and will be assigned to the computer agents when the tasks they are expected to handle become more complex, see discussion in Section 3.3.

3.3 Conversations with Computer Agents

Compared to human-human communication, human-computer interactions have so far been very limited both in subject domain and in communicative tasks. Dialogue technology and industrial applications support fairly well-structured, goal-directed dialogues with well-defined user inputs and system responses. Users who express their needs freely may not get sensible answers, and thus finding solutions to manage misrecognitions, misunderstandings, and other problems in the interaction has been one of the important research goals in developing practical dialogue systems (McTear, 2006). However, in explanation- or information-providing systems, tutoring or coaching systems, the system's communicative capabilities are important.

Moore and Swartout (1992) argued that the problem with dialogue systems is that the response generation is usually seen as a *one-shot process* whereby the system is supposed to give the best and most appropriate response at once. This leads to the over-emphasis of the system's knowledge and reasoning capabilities as it needs to produce an appropriate response in one go without under- or overestimating the user's expertise, etc. In naturally occurring dialogues, however, the users usually ask follow-up questions, elaborations and re-explanations, and thus avoid the "one-shot" generation problem. Moore and Swartout (1992) claimed that a more reactive system is needed: one that understands and monitors the decisions and assumptions made, and with the help of the user's feedback can alter its plans if necessary.

If the system can maintain a dialogue that is both intelligible and appropriate with respect to the user's goals and the application domain, the system's usability is increased. It can also be argued that novice users, who do not know either how the system works or what is the exact coverage of the application, benefit from a system that they can communicate with almost like human partner. With the development of the communicating environment, there will also be an increasing number of interactive applications which the users need to communicate with: they either want to give commands for some particular tasks, or to get to some information. The interactive systems will compete with each other for the user's attention, and expectations for fluent and intuitive communication practises become higher. In order to manage interaction in these situations, communication modelling is needed.

When making the interaction between humans and computer agents more efficient and convenient the system design needs to pay attention to the activities that the humans and computer agents are engaged in, and to the requirements and enablements at the different levels of communication. Natural communicative principles can thus also be extended to cover flexibility in the system's dialogue capabilities. By modelling the system as a rational cooperative agent, it is possible to model the interaction in a manner that better accords with human requirements, and to develop technology that allows implementation of the communicative principles and emergence of cooperative, rational action in the actual interactive situations.

Although the rationality of the system actions may be different from those of the user, the principles of Ideal Cooperation provide the basic assumptions according to which the communicators will behave, and thus function as the basic principles of successful communication also in human-machine interactions. This does not mean that the dialogue agent should not exhibit causal relationship between different input conditions and possible output behaviours, especially when considering such social level communication as politeness codes and interpretation of nonverbal signs (i.e. it is not necessary that all the computer agents always implement reasoning about the ACA principles). Rather, it means that when the agent tries to enable more natural communication with the user, it should demonstrate capability to recognize changes in the input patterns so as to interpret these changes in a relevant manner with respect to its own goals and intentions as well as to the assumed goals and intentions of the partner. The emerging relations can exhibit similar types of systematic relationships that apply to higher-level activities on human psychological and social levels of communication and which are governed by the general communicative principles. To be able to pinpoint the emergence of these relationships and generalisations would, of course, allow us to construct models for robust and intelligent interaction. We call this kind of dialogue modelling *constructive*, as it refers to the construction of communication from the basic principles. We assume that the communicative actions can be formulated using a few underlying principles only, and the application of these principles in a recursive manner then allows construction of the more complex communicative situations.

The Constructive Dialogue Model (CDM) is an approach to communication management between dialogue agents which has the following properties:

1. The agents are rational actors, engaged in cooperative activity.
2. The agents have partial knowledge of the situation, application and world.
3. The agents conduct a dialogue in order to achieve an underlying goal.
4. The agents construct shared context.
5. The agents exchange new information related to the shared goal.

Applied to human-computer interaction, the starting point in CDM is the user with a need or desire which the system tries to satisfy as well as it can. The "object of desire" can be to complete a task or achieve some task-related information but it can also be related to more social aspects of communication like entertainment, or maintaining the social bond. The system, as a rational agent is to address the observed user needs and construct a model of the same "object of desire" that the user aspires for. Consequently, interaction management is a way of controlling the achievement of this goal: it ensures the optimal functioning of the communication as an instrument for achieving the purposes. It must be noticed that optimality does not always mean "efficiency" as is often assumed when evaluating service agents. Instead, it means optimal in meeting the user's needs which may not only deal with accurate information exchange but also of friendship, emotions, and feelings. The evaluation criteria for the optimal interaction can thus be derived from the "object of desire" and the types of communicative activities that the "object of desire" presupposes in order to be achieved.

On the other hand, the "object of desire" is an imaginary goal in the sense that the participants cannot be totally sure if their view of the shared goal is correct. Its correctness is tested in each contribution, and the partners rely on the principles of Ideal Cooperation to converge on the shared understanding. As the dialogue reveals itself, the previous context imposes more and stronger constraints on the possible inferences that the partners can draw. The convergence of this process leads to a closing of the interaction, presumably with a successfull achievement of the "object of desire".

In the following chapters we study how aspects of ACA can be incorporated in a dialogue manager and how the system's general communicative principles can be modelled. As emphasised earlier in Section 3.1, it is not necessarily desirable or even possible to formalise the whole of ACA in order to build simple service systems. For instance, voluntary action and motivation to seek pleasure and avoid pain may not have any real meaning in the computer agent's interactions, while some factors that affect the agent's communicative competence are constant and the agent need not explicitly reason about them (e.g. the roles of the user and the computer agent as an information seeker and information provider). Communicative competence can also appear in the inferences that the system draws in order to provide appropriate responses. This is especially true of rational cooperation: it is not modelled by a single rule which explicitly encodes the system's desired behaviour, but rather, it emerges from the global system design during the reasoning processes. The balance between what should be stored as rules that govern the system's inference processes and what should be perceived by a robust and helpful system as a whole, is a decision that depends on the level of generality that the system aims at. However, it is important to have a general theory of communication, in order to decide which factors of cooperative communication are relevant and need to be taken into account on any chosen level of generality.

Finally, as already mentioned in section 3.2.2, if computer agents' communicative competence is increased, also their roles can become more varied, and consequently, their communicative rights and obligations should grow to match the increased capabilities. To make the computer agent operate more flexibly, the agent may be given the right to initiate topics and conversations, as this is considered useful in resolving problematic situations and misunderstandings. It may also be given the right to make decisions on behalf of the user if these concern specific expert knowledge, e.g. updating software or fixing some computer network problem. While these are basically design decisions that deal with the questions of how much initiative the agent can have, and what are its decision making abilities, they also increase the agent's perceived intelligence: an intelligent partner is likely to have some independence concerning its actions. On the other hand, the increase in the agent's autonomy and decision making ability also brings in considerations of the possibility of the agent's less benevolent and human-friendly actions.

However, it is through the above kind of design decisions that the computer agents' behaviour is constructed in the first place. When initiative and decision making are related to activities that are useful for the human user and can free her from boring tasks, they are usually considered good. The same capabilities combined with a cold rationality that only deals with efficient planning and maximal benefits are likely to be evaluated as bad, creating fear and contributing to the threatening visions of machines taking over the human race. In this book, we try to argue that one solution to building intelligent and "nice" computer agents, and thus balancing the dichotomy of beneficial and destructive promises and threats created by scientific advances, is to teach the computer agents to communicate in a cooperative and constructive way.

Computer agents may develop towards autonomous agents i.e. their behaviour can be determined by their own experience rather than on knowledge of the environment that has been built-in by the designer (definition in Russel and Norvig, 2003). However, in the beginning, with little or no experience, the agent will act randomly unless the designer gave it some assistance. As we seem to be in the beginning phase of developing such autonomous agents, it seems logical that we, as designers of the interaction capability of the agents, also help them to learn communication that is regarded as natural, rational and cooperative by the human standards. Rather than leaving this to be learnt by a random learning process, which may terminate in some undesirable non-optimal local maximum, it seems reasonable that we give the learning agent a push by modelling cooperative communication, based on the assumptions of rational agents, and by designing the agent's communication capability on the principles of Ideal Cooperation from the very beginning.

4

Construction of Dialogue and Domain Information

Rational agents do not act randomly, and their communicative contributions are not random either: there are relevant connections between the utterances and what the agents consider as the topic of the conversation. The CDM approach assumes that the agents construct the dialogue as they go along: they present information, react to the presented information, and the reaction then serves as a new presentation, so that the dialogue is more like a chain of presentation acts than a structured set of actions (cf. Clark & Schaefer, 1989). In this chapter we will have a closer look at this kind of construction process: how the agents control information flow in a cooperative way, and how the construction of dialogue is managed dynamically at each contribution. We will first review two related concepts: coherence, or the aboutness of the dialogue, and new information, or the main message that the dialogue will get through.

4.1 Coherence and Context – Aboutness

In some intuitive sense, coherence means that the dialogue "hangs together" (Stenström, 1994): contributions are linked to each other and the dialogue talks about the same topic. On the surface level, individual sentences and dialogue contributions may concern unrelated objects and events, but as a whole, contributions are linked together into a coherent communicative whole.

In general, there are two types of links: those that hold between ideas, propositions and activities, and those that hold between discourse objects. The difference is often referred to as global and local coherence (Grosz & Sidner, 1986; Grosz, Joshi & Weinstein, 1995): global coherence describes the ways in which larger segments of discourse relate to each other, and thus deals with the overall well-formedness of the dialogues, while local coherence operates on more local focusing processes, and concerns individual sentences and their combination into larger discourse segments. Global coherence is usually captured in terms of task

Constructive Dialogue Modelling   Kristiina Jokinen
© 2009 John Wiley & Sons, Ltd

structure and dialogue plans, whereas local coherence has been captured with the help of salient discourse referents.

The tacit assumption is that the dialogue is coherent, if it talks about the same set of entities. However, as already pointed out by Hobbs (1979), this is not enough to guarantee coherence: rather, coherence requires that entities are connected via some general relations that hold between discourse entities. For instance, the following discourse (from Hobbs, 1979):

(1) John took a train from Paris to Istanbul. He likes spinach.

talks about "John" but would be considered an example of incoherent discourse by most readers. Hobbs argues that coherence results from the agent's need to be understood: this drives agents to seek for an appropriate coherence relation, and thus coherence crucially depends on the agent's ability to find a relevant causal or motivating relation between sentences or dialogue contributions. For instance, (1) can appear coherent, if it is known, or can be inferred, that Istanbul is a good place to buy or eat spinach. The second sentence then provides explanation or motivation for John's taking a train to Istanbul.

The amount of reasoning that the speakers need to do in order to determine whether the discourse is coherent or not, is influenced by the context in which the analysis takes place: e.g. the discourse (1) may be odd in isolation, but the reader can be directed towards the intended interpretation, if the context is extended by a title like "Travels in Turkey – the world's best spinach dishes", and a clause like "and he was eager to try out spinach dishes that his friend called the best in the world" at the end.

Hobbs (1979) further argues that coherence depends on the number and overlap of connections between the text portions, and that the connections themselves can

Figure 4.1 Misunderstandings in interaction. From the Interact-project poster (Jokinen et al. 2002)

hold between sentential units to a greater or lesser degree, depending on how salient the axioms were that were used to establish the relation. Moreover, the salience of the axioms is not constant across all agents, and is not necessarily constant even in the same discourse, but depends on the agents' knowledge about what counts as a possible domain relation. Consequently, coherence relations are tied to agent modelling, the agent's knowledge of the domain and beliefs about the partner's knowledge about the domain. The same discourse may appear more coherent to some agents than it does to others, because their background knowledge is different.

Coherence has been studied extensively in text linguistics and text generation, where the work has especially centred on the conditions and requirements for producing coherent texts and on organising content into coherent paragraphs. Various domain independent coherence relations have been proposed although it is also pointed out that some coherence relations can be more appropriate in some domains than in others: e.g. motivating relations seem more suitable when talking about the benefits of physical exercise for health than when describing bus timetables for a visitor. Domain also restricts the appropriateness of the relations for expressing domain information: e.g. sequencing requires knowledge of what domain facts are appropriate to be included in the sequencing relation.

In dialogue research, the coherence is divided into global and local coherence (Grosz & Sidner, 1986; Grosz, Joshi, & Weinstein, 1995) with the organisation of dialogues into turns and sequences. Local coherence operates via such mechanisms as turn-taking and adjacency pairs, while globally dialogues are organised into sequences of turns, which organise conversations into a specific kind of information exchange. Compared with monologue texts, coherence in dialogues is further complicated by the fact that it is not controlled by one agent, but built collaboratively by the participants. For instance, Moore & Paris (1993) showed that an expert system cannot respond appropriately to the user's follow-up questions if it relies on content-based rhetorical relations only. In addition, knowledge about the user's intentions and goals is required, since the same rhetorical relation can be used for different communicative purposes and the same communicative goal can be achieved via different rhetorical relations. This is basically the same argument that was pointed out as being the main reason for the failure of the Eliza-type dialogues (Section 2.1.1): if the partner only follows the content but does not seem to have any intentions as to tie the conversation together, the interaction is not considered rational but incoherent.

Bottom-up approaches to dialogue modelling emphasise that topics are introduced and developed in dialogues through the subtle processing of the content of the utterance, and coherence is thus constructed in collaboration with the partner across turns (Levinson, 1983; Heritage, 1984). Of course, in dialogue situations, the participants are also bound by the particular topic related to the context of a previously defined task goal.

Coherence serves the participants in their attempts to achieve their task goals and understand the partner's goals in an efficient way, but rather than being the reason for the agents' behaviour, it is considered a side effect of the agent following rational reasoning. Following Hobbs (1979), we can say that coherence is due to the communicators' desire to make sense of the discourse, but we can also make a stronger claim, namely that the communicators are obliged to produce coherent discourse, given that they act, or want to be seen acting, as rational, cooperative agents. Hence, coherence arises from the agents' compliance with the principles of Ideal Cooperation. In order to achieve the underlying dialogue goal, the agents introduce new topics, agree to discuss some topic, and close a topic, and all these activities involve decisions that are governed by communicative obligations that guide rational and motivated behaviour: the agents are obliged to evaluate the partner's contribution in the context, convey the resulting new joint purpose to the partner, and choose a view-point from which the new information is presented in a way that allows the agents to achieve their goals and does not prevent the partner achieving their goals. Because the agents act in a way that they believe is rational in a given context, their communicative actions produce behaviour that they believe is coherent and in accordance with communicative principles. Moreover, because the agents assume that the partners share the same rationality considerations with them, they also believe that their reactions will be considered to be coherent by the partners.

4.2   Information Structure of Utterances – New and Old Information

Kruiff-Korbayova & Sreedman (2003) provide an excellent overview of the definitions and history of the terminology concerning such concepts as topic, focus, old and new information, foreground and background, which are related to the flow of information in the discourse. Detailed distinctions have not always been made in the computationally oriented literature, and the concepts topic and focus have also been used somewhat differently, especially the focus being associated with the centre of attention, and with salient entities which link utterances together and can be pronominalised, i.e. with aboutness of dialogues, rather than with newness of discourse entities.

The distinction between new and old information is encoded in the information structure of the sentence. The word order often marks the difference so that the theme or topic which contains old information occurs at the beginning of sentence, while new information is usually presented at the end of the sentence, and is thus maximally distinguished maximally from the topic. In languages which have a free constituent order (e.g. Finnish), word order marks the information status of the constituents rather than their syntactic functions, i.e. certain positions in the sentence structure indicate the topic, the new information, and the possible contrasted element rather than the constituent functioning as a subject or an object (which functions are marked by inflection or by certain particles). The interpretation of the constituent's information status is thus guided by discourse configuration: the presence of certain discourse patterns directs the hearer's expectations concerning the upcoming elements (Jokinen, 2005c). Also, cognitive studies (Kaiser & Trueswell, 2004) show that hearers make efficient use of the non-canonical word order patterns in order to predict upcoming referents and their discourse status. The speakers' awareness of the information structure is thus part of their communicative competence.

Information structure also has distinct prosodic marking: information units are marked by speaking rate, pitch range, loudness, duration of segments and boundary tones. Intonational characteristics of topic and focus (old and new information) have been much investigated (see examples and discussion in Section 2.1.3.1). In general, new information is usually accented, and the given information de-accented (Table 4.1 ex. 1). This need not always be the case, and old information can be accented, if there is a change of topic (Table 4.1 ex. 2) or items are contrasted (Table 4.1 ex. 3). Nakatani (1996) points out that new and old information are accented when the speaker's focus of attention shifts to the item. For instance, pronouns can be accented in order to keep a backward looking focus element in focus, or to bring a non-focused given referent into focus, and given information is accented when brought back to attention after a digression or if it has not been in focus. Listeners expect certain parts of an utterance to be in focus and marked prosodically: new information, items to be contrasted with something that is discussed earlier, and changes of focus are important aspects of the message that need to be easily distinguished. In some specific cases, the whole meaning can vary depending on the intonation (Table 4.1 ex. 4).

Another aspect of information structuring is contrast. This is orthogonal to the old-new dichotomy of an utterance (Vallduví & Vilkuna, 1998). A contrasted element is the main focus of the utterance but it can appear anywhere in the sentence, being marked by intonation or focus-markers (*only, even*), and in some languages, like Finnish, it can be marked especially by a word order. Confusion can occur because the contrasted element can be new or old; in particular, it need

Table 4.1 Examples of old information being prosodically marked. Accented words are written with capital letters. Last example from Halliday (1970a)

Example 1	*When does the last bus leave?*
	The last bus leaves at ELEVEN O FIVE pm
Example 2	*The last train leaves at 11:35 pm*
	And the last bus?
	The last BUS leaves at 11:05 pm
Example 3	*And the last bus arrives at 11:05 pm*
	No, the last bus LEAVES at 11:05 pm
Example 4	*DOGS must be carried – you must carry a dog*
	Dogs must be CARRIED – carry your dog

not contain discourse new information but may have been discussed earlier in the dialogue. In contrast, the whole utterance is new information although the discourse elements as such may be old: the new information in contrast is the contrast itself. The contrast thus transcends the usually entity-related information structure into a propositional world where the information flow concerns ideas, events and plans rather than participants in these events. The information structure must thus deal with propositions and ideas besides objects as such.

Approaches like the Centering Theory (Grosz et al. 1983, 1995; Walker et al., 1998; Hahn and Strube, 1999) and Information Packaging (Vallduví, 1995; Vallduví & Engdahl, 1996) provide structured accounts of discourse referents and their role in information structure.

Approaches to information structuring often concentrate on the surface structure of a single utterance, and when applied to dialogues they tend to underestimate contextual information as well as the goal-oriented constructive nature of interaction: utterances are not produced in isolation but as a reaction to what has been previously talked about. A more discourse-oriented way to define topic is found, e.g. in van Kuppevelt (1995) and Vallduvi (1995) who suggest that aboutness is defined with the help of questions that the contribution or discourse addresses. In a coherent discourse, each sentence constitutes an answer to an implicit or explicit question which has been raised in the preceding discourse. The questions describe what the discourse is about by reconstructing the appropriate context, and new information can be defined with respect to the given topic as being the new material that answers the question. For example, in Table 4.2, the first question defines the topic, i.e. the central concept "Lisa", while the second question specifies the new information that is presented about the topic. Depending on the question, the new information can concern the action or activity Lisa participates in, e.g. buying an ice-cream as in the case (a), or an aspect of her action such as the object ice-cream in the case (b). In both cases, the thematic structure of the utterance in terms of topic and comment is the same, but the information structure in terms of new and old (focus and background) information differs.

Table 4.2 Implicit Questions and Information Structure of utterances

(a) What about Lisa? What did she do?

She	*bought an ice-cream.*
Topic/Theme/Link	Comment/Rheme
Old/Given/Background	**NewInfo**/Focus

(b) What about Lisa? What did she buy?

She	*bought an ice-cream.*
Topic/Theme/Link	Comment/Rheme
Old/Given/Background	**NewInfo**/Focus

In question-answering dialogues, the explicitly posed requests function as topical questions, although it can also be argued that there must be other implicit questions that can be used to reconstruct the context for the overt questions and of which the overt questions are "sub-questions"; in other words, there are inferential links between the questions not overtly expressed in the utterances.

This kind of aboutness defines the topic of a contribution as well as that of the whole discourse. Implicit or topical questions are analogous to coherence relations discussed in the previous section since they also link propositions in consecutive sentences together. However, implicit questions differ from the coherence relations in that they are closely related to the domain and the agent's intentions. Implicit questions and the aboutness within the whole discourse can be compared with the discourse theory of Grosz & Sidner (1986), where the discourse is segmented according to discourse purposes, i.e. the goals that the agents try to achieve by communicating, and which can dominate each other so as to form nested discourse structures. The implicit topical questions can be understood as rephrasing these goals on the content level.

Of course, it is necessary to supply a mechanism in order to determine whether and how implicit questions are related to each other, and what kind of implicit questions can there be in the first place. If the questions are reconstructed backwards on the basis of the actual answers as in Table 4.2, the process conflates with interpretation and resembles abductive reasoning (Hobbs, 1979). The interpreter may not end up with the "correct" interpretation and "correct" topical questions, but with something that satisfies the coherence and consistency of the discourse from her point of view. The agents are likely to share the general knowledge framework, and thus the result of the inference is usually the intended one.

The implicit questions are linked to each other with the help of rational considerations. The agent tries to find appropriate questions in analysing the partner's contribution in the context of her own expectations that guide and restrict her reasoning. The aim is to find questions that would link the utterance to the previous discourse, and motivate the partner's contribution. Usually this is quite easy since the partner behaves according to expectations: e.g. the partner may answer the agent's question, or communicate progress of some task. In other cases it requires more intensive reasoning: the coherence of what the partner said and the previous context is not immediately obvious.

It is clear that the relevant topical questions need not be the same for the agents, and their rational considerations need not always get across to the partner. For instance, Hobbs (1979) argued that coherence is a continuum that does not only depend on the agent's knowledge about the domain, but also on the salience of the axioms that were used to establish coherence reaction. Misunderstandings in the agents' reasoning because of the discrepancies in the salience of such axioms can be exemplified by the dialogue in Table 4.3 where the problem concerns a misunderstanding of the focus of attention and contextual reference: what is it

Table 4.3 Example dialogue of misunderstood referent (from Jokinen and Wilcock 2006)

F:	Have you seen *the new curriculum plan*?
E:	Yes, and I'm annoyed because *my first year computation course* has been moved to the spring term.
F:	**It** is really a bit hastily put together, I agree.
E:	No, my course is well-planned, certainly not put together hastily.

that we are talking about? The referent for the pronoun "it" is misinterpreted, and the result is potentially offending.

In syntactically oriented languages such as English, anaphoric pronouns usually have a "narrow" scope i.e. the preferred antecedent is to be found among the elements mentioned in the immediately preceding utterance (cf. Centring Theory discussed above). The appropriate referent for the bold-faced *it* is thus *my first year computation course* which, however, receives an unacceptable and confusing interpretation with the rest of the utterance. In discourse oriented languages like Finnish, by contrast, the anaphoric context is the discourse situation itself, containing the topic and the discourse referents which have been introduced earlier in the discourse. In the case above, the intended referent for *it* is *the new curriculum plan*, since this was the topic introduced by the speaker in the beginning of the interaction, and to which the partner's contribution was understood as providing new information. In other words, *it* continues the original discourse topic *the new curriculum plan*, since there is no indication that a topic change is present.

4.3 Definitions of NewInfo and Topic

In order to capture aboutness and information flow in conversations, we distinguish two interrelated dimensions in the dialogue: the relevance of information and the salience of information. The former concerns the thematic structure of contributions, i.e. the content of the contribution in terms of what the contribution is about and what is background, while the latter encodes the information structure of contributions, i.e. old and new information conveyed by the contribution. The two dimensions are modelled with the help of the following two concepts which are relational concepts in that they are defined with respect to the agent's world model (i.e. with respect to the agent's beliefs of the world):

- *Central Concept (CC)* is a distinguished discourse referent which is talked about in the contribution. It represents the *focus of communication*, and sets the view-point from which information is presented, thus narrowing down relevant background assumptions.
- *NewInfo (NI)* introduces a discourse referent which is new with respect to some CC. It represents the focus of attention at a given moment, and encodes the

information that is intended to be exchanged with the partner by the contribution. It can also be described as what Chafe (1976) calls *conscious information*: "what the speaker assumes he is introducing into the addressee's consciousness by what he says".

In practical dialogue systems, CC and NewInfo are often encoded by single discourse referents, but this need not always be the case (and in human conversations it seldom is); there can be several referents which represent the topical information and novel information (cf. Section 5.2.2). The set of referents must fit into human cognition, however, and accord with the limits of the short-term memory; otherwise the agent's processing ability is exceeded. An interesting research question is then, what is the minimal information exchange unit and how does it operate in communication. Following Levelt (1989), we can loosely say that NewInfo must be a minimal intonational unit which has a meaning, takes the form of a word or a phrase, and can be produced within intonational constraints. Concerning the agent's cognition and consciousness, the size of NewInfo is restricted by the capacity of the working memory: the focus of attention usually holds seven plus or minus two items. However, here we continue on a more linguistic level by considering how symbolic representation and language processing can be modelled in interactive systems.

CC and NewInfo have different scopes. While CC is defined with respect to that part of world knowledge which the agent has distinguished as the background context for dialogue contributions in a particular dialogue situation, NewInfo is defined with respect to this context as information which has not yet been mentioned but which the agent intends to bring forward about the CC in the dialogue. The difference in scope captures the fact that, in order to react, a rational agent need not consider the overwhelming amount of world knowledge that is new in a given situation, but can restrict reasoning to the relevant part (see Chapter 3). At the very beginning of the dialogue, when the agent introduces a new topic, the entire proposition is of course new, i.e. NewInfo. In this case, NewInfo can be regarded as being selected with respect to the entire world knowledge: it brings in the part of the knowledge that the agent's attention is focussed on and the agent wants to set as the context, or the background information for the coming dialogue.

As a consequence of the difference in scope, NewInfo is usually chosen within the selected CC: the order reflects the fact that contributions are formed around a focus of communication about which something is intended to be made known to the partner. If the agent cannot find any new aspects of a CC to be brought in the discourse, she can shift the focus of communication onto a new CC and then produce NewInfo about this topic. As a corollary, NewInfo is the information centre of the contribution: it must be present explicitly in the contributions, and it cannot be pronominalised. CC, however, need not be realised explicitly (elliptical utterances), and it can also be referred to by a pronoun. If the speaker's attention changes to a new CC, this must, of course, be introduced explicitly in the dialogue, and the partner must be notified appropriately about the change in the point of view.

NewInfo is always different in consecutive contributions. It is not rational to carry on repeating the same NewInfo, and violations of this cause the hearer to look for a reason for the repetition: a failure in contact, perception or under-standing, some humorous purpose, or the partner simply behaves in a manner which is not quite "normal". Re-phrasings "in other words" are not regarded as repetitions, however, since they do contain new information, namely the new phrasing with another view-point.

4.3.1   Discourse Referents and Information Status

The world model is the agent's private model and it comprises of the information that is known to the agent. The concepts of CC and NewInfo allow us to partition the agent's knowledge about the world and the dialogue situation into different, partially overlapping sections. On one hand, the world model can be partitioned into the relevant background information and the rest, in regard to a particular focus of communication (CC). On the other hand, the background context can be partitioned into information that is new and salient (NewInfo), and into infor-mation that is old and less salient to the agent at a given moment. It should be pointed out, however, that the notions of relevance and salience form a scale rather than a binary classification. The agent's knowledge can be further divided into information which the agents share, and the information which is private to the agents. The former has been introduced and made explicit at some point in the course of the agents' interactions, while the information that is part of the agent's private knowledge is potential NewInfo in the agents' future interactions.

Given that the agent's world knowledge is represented as concepts which can be instantiated as discourse referents, we can say that all discourse entities related to a particular CC form the background information for that topic. They are activated in the dialogue situation, although not necessarily brought into the partner's attention by explicit reference. The conceptual distance of the related entities from the CC determines the relevance of the entities with respect to the CC. The closer to the CC a discourse entity resides, the more relevant it is and the more likely it is that the partner understands its relevance in the dialogue context. Of course, proximity of concepts is not an absolute relation but is conditioned with respect to the agents. Since the agents have different world views, the relations which are relevant for one agent, may be considered clumsy and irrelevant by the other. The concrete basic environment is usually experienced and conceptualised in a similar manner by the agents, but events and abstract concepts form less uniform hierarchies and conceptual networks. The issues in ontological modelling discussed e.g. by Gruber (1993). Assuming that the world knowledge is encoded in the internet, the issues become pertinent also in the currently active research area of the semantic web (see, e.g. McGuinness, 2002 and references there).

It is important to note that the concepts belonging to the background information and to the old information are different although they overlap. While background information is part of the world knowledge that has been activated in the agent's mental model through some CC, old information refers to the information that has been presented in the course of the agents' dialogue interactions, either the current one or some earlier ones. Background information can be either old or new to the listener: it can be something that the agents have already discussed, or it can be known only to one of the agents in which case it can be introduced to the partner in the dialogue as NewInfo. For instance, if the agents are involved with the activities concerning searching and providing bus timetable information, the scenario also activates public transportation as relevant background information, while the previous encounters by the agents (i.e. the information seeker and the information provider) make up old information. The NewInfo can be selected within the activated background information which is not yet part of the (shared) old information, or within the information which is not salient enough and needs to be re-introduced.

We define old information as information that has been discussed in the course of the dialogue; it is previously exchanged NewInfo between the partners. Old information is thus related to some CC, and to the concepts it has activated, but does not necessarily comprise all of the possible background information that the CC activates for the dialogue partner. In particular, some parts of the background information may remain tacit and never become old information even though activated in the agent's mind: it may be that the agent never chooses to talk about them. On the other hand, old information is a time-related concept: it marks the recency of concepts in the agent's consciousness. Thus there are different "degrees" of salience of old information depending on when it was active in the agent's mind and, in particular, it need not originate from the current interaction. The agents may have had a long series of interactions (as is the case not only with friends and family members, colleagues and acquaintances, but also with users and various interactive systems), and old information stems from these earlier encounters. The amount of old information can thus vary in interactions (cf. long-term friends and new acquaintances), and affect the way different topics are introduced and discussed. Of course, some part of the old information may not be relevant in a given dialogue situation, i.e. it is not related to the current CC and not relevant background information. However, the previously constructed shared contexts guide the agent's behaviour: they represent what the agents have learnt from and about each other, and help the agents to align their behaviour in new situations accordingly. Yet another distinction must be made relating to the information status of the discourse referents, namely whether the information about them is known or not known to the agent (i.e. the agent believes she knows the information). The agent must of course have a minimum acquaintance with a concept in order to be able to talk about it, but it is through the interaction that the agent learns more about their environment and can establish new knowledge. With the known/unknown distinction it is possible to draw limits for the background and old information that the agent has in interactive situations: the information the agent can talk about, in other words, the background activated in a particular situation must be known to the agent so she can talk about it. On the other hand, not all knowledge of the agent need be old in a given situation with a particular partner: the agent may possess information that the partner does not know about and which thus is not part of the old information in their shared interaction history. The asymmetry in the agents' knowledge is usually the reason for their communication and a starting point for their exchange of information: the agents can extend their knowledge to new situations and topic domains by expressing their intention to learn more, to get help, or to provide the partner with some new information through the interaction. The intention becomes the joint purpose of the (ideally) cooperating agents, and through their interaction, the agents thus establish mutually grounded knowledge: they mutually come to believe that something is the case. The belief is then established as old knowledge shared by both agents and will influence their later interactions. One of the crucial aspects of establishing shared information is that by so doing the agents also create and strengthen social bonds between them: the shared experience ties the agents together.

It is important to notice that the old and new information, are relative concepts as they are defined with respect to some communicative situation: they depend on the agents' previous interactions and thus the agents adapt and align themselves with respect to each partner separately. The agent's knowledge, however, is defined with respect to the agent alone, and it comprises of all the old knowledge that the agent has achieved through interactions with other agents and with the environment. Obviously, since the autonomous agents have had different interactions, their knowledge of the world differs too. The mismatches between the agents' knowledge usually function as sources of smooth information exchange, but they can also cause misunderstandings and non-understandings. However, assuming that the agents are rational agents and follow the principles of Ideal Cooperation, their communication is aimed at extending their understanding of the environment and at aligning their knowledge with the partner so as to construct shared knowledge. The agents' clarifications and negotiations become part of their shared experience and strengthen their feeling of togetherness, unless of course, the mismatches are so huge that the communication becomes impossible altogether ("no common language"), see Section 3.1.3.

4.3.2 Other Two-dimensional Work

Prince (1981) summarises various definitions for given-new information and proposes a well-known taxonomy of discourse referents based on their assumed familiarity: new, inferable, and evoked. Each type has sub-classes (the examples below are from Prince, 1981): new referents can be either brand-new, for which a referent must be created in the discourse model (*I bought **a beautiful dress***), or Unused, for which a referent is assumed to exist already in the hearer's model although it needs to be re-introduced (***Rotten Rizzo*** *can't have a third term*); Inferable entities can be inferred on the basis of the context (*I went to the post office and **the stupid clerk** couldn't find a stamp*) or from the expression itself (*Have you heard **the incredible claim that the devil speaks English backwards**?*); and the entities can be Evoked either textually (*Susie went to visit her grandmother and **the sweet lady** was making Peking Duck*) or situationally (***Lucky me** just stepped in something*).

Naturally, Prince's Brand-new entities correspond to our NewInfo referents: they cause a new discourse referent to be created in the context model. The Unused, Inferable, and Evoked entities coincide with those activated through the introduction of a particular CC, and in the above example sentences, when brought explicitly to the attention of the listener, they all function as the NewInfo of the introductory presentation sentence. However, some of them may be part of the old information, i.e. something that the agents have talked about before or which belongs to the general world knowledge (e.g. "Rotten Rizzo", "Susie", "the grandmother"), or they may be new to the listener ("Who is Rotten Rizzo; Oh I didn't know that Susie's grandmother is still alive"). In other words, they may introduce brand-new discourse referents into the shared context, or they may re-introduce referents that are assumed to be already known to the listener. The reason for the Unused, Inferable and Evoked entities being familiar is due to them being part of the activated context: they encode salient and highly relevant background information, usually a crucial part of the scheme or a script introduced earlier.

Prince (1981) also emphasises the asymmetry of information between the speaker and the hearer: moreover, she discards the term "shared information" and points out that this is basically the same as the speaker's assumptions of the partner's assumptions; familiarity refers to the speaker's assumptions of what the hearer knows rather than to some overlapping amount of information that both happen to share. Although we use the term "shared knowledge", it must be emphasised that this knowledge is exactly what the agent *believes* the partner believes, i.e. we do not take the position of an omniscient observer but keep the definitions related to the single agent's own view-point (see discussion on communication modelling and minimal rationality in Section 1.2.3). Furthermore, we emphasise the relative nature of new and old information in regard to the interacting participants: the knowledge that is assumed to be shared between two agents need not be shared between the agent and another agent. In multiparty dialogues this is of course one of the problematic cases: the speakers may not be able to adjust their speech according to the knowledge level of all the other partners, and they may not be able to control if and how the information presented during the interaction is grounded in the audience. It is also good to remember that NewInfo and CC are related to the particular communicative situation. For instance, in a discourse such as "The one who uses buses most often in our family is dad", the NewInfo is "dad", although the referent of "dad" is likely to be old in this context, i.e. it is an Unused referent.

The two independent but interacting levels to describe information flow have also been proposed in linguistics. For instance, Steedman (1991, 2000) proposes a similar distinction, although with different terminology, and also Halliday et al. (1967, 1970b, 1985) distinguish the information structure and the thematic structure of sentences. For Halliday, the former deals with information units and the latter with their linear ordering and text-organisation strategies. The information structure is hearer-oriented and it concerns Given (i.e. optional information that is recoverable from the context) and New (obligatory and non-recoverable information), while the speaker-oriented thematic structure divides the sentence into a Theme (point of departure) and Rheme (what is said about the Theme). The definitions for CC and NewInfo are similar to Halliday's Theme and New but extend them to cover the whole dialogue and interaction history.

In dialogue systems, the distinction between the CC and NewInfo has not usually been singled out explicitly. The importance of Central Concept is widely acknowledged, but NewInfo is often implicit: it is tacitly included in dialogue scripts as successive steps of the execution of a task plan. However, in the discourse theory by Grosz and Sidner (1986) and the dialogue systems based on it (e.g. Collagen, Sidner et al., 2000), the two levels of discourse management are defined as the intentional structure and the attentional state, encoding the speakers' intentions and their focus of attention. The former deals with discourse segment purposes which form discourse structures according to the so called Immediate Dominance and Linear Precedence rules, while the latter encodes the focus space which consists of a particular discourse segment at a time and takes care of the discourse referents that are in the speaker's focus of attention. The two levels are analogous to the distinction which we make with the notions of CC and NewInfo: dialogue around one particular CC can be regarded as a discourse segment which defines focus space within which the NewInfo, or the focus of attention can operate. On the other hand, CC also singles out relevant background information from the general world knowledge, while NewInfo distinguishes what is new and old in the communicative situation and helps to construct shared knowledge of the communicative situation by introducing salient information in the listener's mental model.

Especially, in mobile contexts, where the focus of attention can change dynamically as something new takes place in the environment, it is important to distinguish between the newness and topicality of information: the agent cannot rely only on a predefined task structure in order to process the partner's contributions but needs to cope with a changing context and the partner's various objects of interest. The status of the exchanged information is thus deployed in constructing continuation from one utterance to another as well as in distinguishing background information related to a certain topic from the shared information that the speakers have constructed in their earlier interactions.

4.4 Topic Shifting

In the beginning of a conversation, as mentioned above, the context is usually open and the whole utterance is NewInfo. Unless the intended topic is clearly marked, the listener has some freedom to select one of the NewInfo concepts as the next topic she would like to continue the dialogue on. If this is not the chosen topic of the partner, some negotiation about the topic may occur among the participants (Heritage, 1984). As shown in Table 4.4, the evaluation of NewInfo can result in, e.g. requesting further clarification (a) or providing evaluative statements (b–d). If the speaker has singled out a special topic, a related NewInfo can be understood as an explanation or justification of a new request (e).

Later in the conversation, topic shifting is more constrained, since the previously constructed shared context restricts continuations which will be considered coherent. A common way to deal with topic shifting is to use a stack mechanism. (In the earlier approaches, the term *focus stack* was used). The stack keeps the most recent topic (or the focus) as the most accessible one by pushing it into the stack and popping it off again after it has been discussed. The dialogue is considered finished if the stack is empty. Besides single concepts, the stack may also contain more elaborated structures in terms of "context spaces" (Reichman, 1985) or "discourse segments" (Grosz & Sidner, 1986). In the more structured stacks, each stack element contains a list of concepts which can be ordered heuristically so as to mark the most likely next continuations according to their saliency.

Table 4.4 Central Concept and NewInfo at the beginning of some dialogues. Semantic predicates are meant to illustrate the dynamics of CC and NI, not the actual representation: eg: 'car(c)' refers to the speaker A's car ("my car"), and intentions (such as *know, intend to know*) are not part of information structure

	Utterance	CC	NI
a	A: *my car is broken* B: *how do you go to work then?*	broken(c), car(c) broken(c), car(c)	broken(c), car(c) get(work)
b	A: *my car is broken* B: *oh dear, not again!*	broken(c), car(c) broken(c)	broken(c), car(c) again
c	A: *my car is broken* B: *your car is too old*	broken(c), car(c) car(c)	broken(c), car(c) old(c)
d	A: *my car is broken* B: *you better use public transportation anyway*	broken(c), car(c) transportation	broken(c), car(c) transportation
e	A: *I'd like to know about bus timetables.* *My car is broken.* B: *Sure, which bus do you want to know about?*	bus_info broken(c), car(c) bus_info	bus_info broken(c), car(c) which(bus)

The stack is somewhat inflexible in topic management, however, since it requires the pushing and popping of topics in a particular order, and once the topic has been popped from the stack, it cannot be referred to except by re-introducing it. McCoy and Cheng (1990) thus suggested a tree-structured discourse model, a "focus tree", which allows for traversal of the branches in different orders, and thus supports a more flexible management of topic shifts. Recently, mobile contexts have also brought in a need for more flexible topic management: the user's focus of attention changes depending on the dynamic context, and the topic tree can accommodate the agent's observations better than a simple stack (cf. Beun 1998; Hinze & Buchanan, 2005).

McCoy and Cheng (1990) define the focus tree, or according to our terminology, the topic tree, as a sub-model of the conceptual structure of the domain of discourse. The tree is a network of related concepts which represent the agent's knowledge about a particular topic: it picks out a sub-graph of the agent's world model and represents what the agent considers as the focused concepts of the discourse. (In the context of the internet, the conceptual model can be from the semantically annotated Semantic Web, (in which case the computer agent's knowledge is limited with respect to the results of the search engine.)

Different node types, such as "object", "attribute", "setting", "action" and "event" (McCoy & Cheng, 1990; Bateman, 1995; etc.) or various metadata tags and schemas in the Semantic Web allow for different shifts and help the agent to determine what sort of shifts are likely to occur, and whether or not a shift is easy to process. Without going into the details of possible conceptual relations or building of ontologies in general (see, e.g. Gruber, 1993), we simply assume that the world model exhibits relations that link concepts in a way that represents the agent's coherent world view. For instance, the ontology should allow the agent to talk about properties of an object, participants in an event, and settings of an event. Events can be further divided into states and actions, objects can have different sub-types, and the model can also include a special concept of a (rational) agent, to be instantiated either as a user or a system.

With the help of the topic tree, dialogue coherence can be defined on the basis of the types of relationships that occur in the tree: the dialogue is coherent because the topic nodes are related to each other through the coherence and consistency of the underlying conceptual model. Topic shifting is then defined with respect to possible relations among the various kinds of world model concepts. The following three types of topic shifts can be defined:

- A *coherent* topic shift is a shift which follows the coherence relations that exist in the agent's conceptual model. McCoy & Cheng (1990) assume that a coherent shift from an object-type concept is to the properties of the object, to the event the object participates in, or to a sub-class or super class of the object. Coherent shifts from an event-type concept are usually to the participants of the event, to the setting (location and time) of the event, to the next action in sequence, or to a sub-class or super class of the event, and coherent shifts from a setting-type concept are to the event that takes place in the setting or to a sub-class or super class of the setting concept.

- An *awkward* topic shift is a shift which is not coherent, but the current topic and the previous one can be found to be related thematically to each other. Thematic relatedness in general means inference links between the two topic nodes, and the distance between the nodes should be less than some limit (the smaller the limit, the stricter the agent is in considering what is relevant in the dialogue). Since relatedness depends on the agent's world model, a cooperative agent should mark awkward shifts by a discourse marker (by the way, then, going back to, etc.), and preferably also explicitly close the previous topic (ok, fine, etc.), so that the partner would find it easier to evaluate the linking between the topics.
- An *incoherent* topic shift is a shift which is not awkward. Usually this means that the agent has initiated a brand new topic.

4.5 Information Management as Feedback Giving Activity

NewInfo is the information centre of a contribution which the agent reacts to. The construction of the shared knowledge starts with the analysis and interpretation of NewInfo, and the agent has two tasks to perform: to evaluate the relevance of NewInfo in the shared context, and to consider if it is an acceptable contribution to the joint purpose, then report back the result of the evaluation to the partner. Acceptance and reporting are important in the construction of the shared knowledge-base, since they make up the process of grounding.

Following Allwood (1976), we assume that the agent's reaction includes evaluation of the basic communicative functions: contact, perception, understanding, and attitudinal reaction, derived from the basic requirements for communication (see Chapter 3). The evaluation of NewInfo is a crucial part of the inferences on the level of understanding, i.e. assessing the relevance of the information for the assumed goal of the interaction and whether or not to accept the content as part of the shared context. The agent's reaction is then determined through evaluating how to report the result of the previous evaluation processes to the partner. Although evaluation of the basic functions is usually simultaneous, the functions logically form a hierarchy, so that a failure on one function means that the evaluation of the higher ones cannot succeed either. For instance, understanding presupposes perception which presupposes contact, and so, if the agent is in contact with the partner but cannot perceive any, there cannot be any evaluation on the level of understanding, either, and the agent needs to report this to the partner. In what follows, we often talk about evaluation of NewInfo, but it is good to remember that this always presupposes that the basic contact and perception are also evaluated successfully.

From the agent's point of view, communication consists of an analysis of the partner's contribution, its evaluation with regard to the agent's own knowledge and intentions, and reporting the evaluation results back to the partner as the agent's reaction. The process can be compared with what Clark and Schaefer (1989) call the presentation-acceptance cycle of communication: the speakers are engaged in an activity through which the agent presents information which the partner accepts (rejects), and the partner's acceptance (rejection) functions as a new presentation which the speaker in turn has to accept (or reject), etc. Although there are questions about how to express acceptance and rejection and how to finish the cycle (see e.g. Traum, 1999 for a thorough analysis of the presentation-acceptance cycle), the main point of the communication being a continuous activity of information exchange is the same as that which CDM advocates. However, presentation and acceptance are performed by different agents, whereas the analysis-evaluation-reporting cycle describes processing from the view-point of a single agent only. This is not a big difference, and in fact CDM can be seen as a more detailed description of Clark and Schaefer's acceptance phase or, if we consider the reasoning processes that result in the presentation of information, of the presentation phase. The cooperation that is built in the presentation-acceptance cycle is encoded in a detailed way into the agents' communicative obligations. However, what is emphasised in the CDM is the agent's view-point: rather than looking at the communication from the omniscient closed world perspective, communication is studied from the perspective of one single agent, possessing limited knowledge of the world and other agents, and trying to coordinate actions and collaborate with the partners in accordance with general communicative obligations. The exchange of NewInfo produces changes in the communicative context, and both the speaker and the hearer are obliged to evaluate NewInfo in order to manage the contextual changes. Although NewInfo is the same for both the speaker and the hearer, it is evaluated differently: the speaker considers how well NewInfo carries the content she intends to contribute to the construction of the shared knowledge, while the partner evaluates NewInfo with regard to how well it fulfils what she expects to be an appropriate contribution to the shared knowledge. The differences in the agents' viewpoint and knowledge then drive the communication further towards a shared understanding and a shared view of the communicative situation.

While the evaluation of new information is motivated by the agent's epistemic needs to cope with the changed context, the same does not hold with the reporting of the results of this evaluation: a selfish agent could simply stop after evaluation without any particular need to respond. This is where ethical considerations come in: as the agent is engaged with the partner in a coordinated activity, she needs to consider the partner as well. In other words, she needs to inform the partner of her understanding of the new situation, so as to allow the partner to work towards the shared goal with all the relevant information. The agent's decision to report the evaluation is thus based on her complying with the requirements of ethical considerations. Consequently, if the agent does not respond, she is regarded as extremely rude, unpleasant and in violation of the principles of Ideal

Cooperation, unless there can be other explanations such as the agent being angry or not hearing, etc.

The partner's reaction functions as immediate feedback to the agent about the success of her contribution: it indicates if the partner has understood the intended meaning and accepted it as part of the context. Using the terminology from the Speech Act Theory, we can say that the partner's reaction shows how *felicitous* the agent's contribution has been in the context, and in fulfilling the partner's expectations and assumptions about the communicative situation. It must be emphasised that the notion of feedback that we talk about here has a wide meaning that describes the evaluation process in the communication cycle in general, and it differs from the more narrow sense of an explicit speech act also called feedback or confirmation. While the agent can choose speech acts, and e.g. decide whether to give explicit or implicit confirmation or no confirmation at all, it is not possible to choose whether to give feedback or not: one's reaction always provides feedback to the partner.

The relations between the evaluation of NewInfo and the different types of reactions as feedback giving activity are shown in Table 4.5. The basic enablements of communication are considered to be conditions that may be true, false or only partially realised and, based on the evaluation results, the agent's reaction can be classified into different feedback types. The classification thus extends the work by Allwood, Ahlsen and Nivre (1992) who also relate feedback signals to the evaluation of the basic enablements of communication, and regard feedback as one of the main ways in which cooperation is pursued in dialogue. While their focus is in the semantics and pragmatics of linguistic feedback in human-human conversations, we define various the conceptual possibilities of various NewInfo evaluations for the purposes of computational dialogue management.

Feedback is often accompanied by non-verbal communicative signs like gestures, facial expressions, and gaze, which usually are intuitive and unintentional. Drawing on the semiotic terminology of Peirce (1960), Allwood (2002) points out that the speakers "indicate" the non-verbal feedback signs, but do not consciously present or "signal", them. However, explicit non-verbal feedback can also function as a deliberately chosen means of presentation (e.g. putting a finger to one's mouth to signal "silence", or raising one's eye-brows to signal surprise). Some aspects of feedback expressed by hand gestures, facial expressions and body posture are discussed in Allwood et al. (2007). Multimodal feedback is included in our classification in Table 4.5, although we will not go into detail here.

Table 4.5 shows the combinations of contact (C), perception (P), and the agent's beliefs about whether she has understood the partner's contribution (Understood), whether NewInfo is relevant in the context (Relevant), and whether the agent can accept NewInfo as part of the shared knowledge (Accepted). All examples are possible responses to the utterance *"I'd like to go to Eira"*.

Given that the basic contact and perception are ok, the agent faces three large evaluation tasks: she first needs to produce a meaningful semantic representation

Table 4.5 Evaluation of NewInfo and feedback types. All examples are possible responses to the utterance *I'd like to go to Eira*

Reaction type	CP	Understood	Relevant	Accepted	Example responses to *I'd like to go to Eira*
Signal CPU	CP	Yes	Yes	Yes	When?/How?/Why? That's a nice idea!
Signal unclarity	CP	Yes	Vague	Yes	Please clarify why do you say that?
Signal irrelevance	CP	Yes	No	Yes	How is this relevant here?
Signal need for confirmation	CP	Partial	Yes	Yes	When/how/why would you like to go to Eira?
Signal need for clarification	CP	Partial	Vague	Yes	Did you say: I'd like to go to Eira?
Signal need for specification	CP	Partial	No	Yes	What do you mean by Eira?
Signal explicitly non-understanding	CP	No	–	Yes	Sorry, I don't understand what you mean
Signal implicitly non-understanding	CP	No	–	Yes	I see, when will it start?
Signal problem in perception	C	–	–	Yes	Sorry I didn't hear you, what did you say?
Signal conflict	CP	Yes	Yes	No	That's not a good idea
Signal doubt	CP	Yes	Vague	No	I wonder if it is a good idea
Signal reject	CP	Yes	No	No	You didn't understand
Signal non-acceptance	CP	Partial	Yes/Vague	No	You cannot go to Eira
Signal objection	CP	Partial	No	No	That's not a place for you
Signal ignorance	CP	No	–	No	I was saying ... /<ignore>
Signal unnoticed	C	–	–	No	I'm busy/<ignore>
Back-channelling	CP	–	–	Yes	uhum
Non-verbal	CP	Yes	Yes	Yes	<raise eye-brows>

for NewInfo, then decide whether it is coherent or thematically related to the current topic, and finally whether it addresses the evocative expectations of the agent's previous contribution. If everything is fine, the dialogue continues smoothly, and the agent can choose a suitable reaction from a large repertoire of different reaction types at her disposal. However, even though NewInfo is evaluated as vague or totally irrelevant in the context, the agent can still accept the contribution and continue the dialogue by asking the partner to explain the relation (signal unclarity) or by expressing explicitly that NewInfo lacks relevant links (signal irrelevance). In this way the agent signals need for more information to the partner but the dialogue can still continue smoothly as a negotiation rather than come to a halt because of a problem.

In order for the agent to reason about the relevance and acceptance of the contribution, she must be aware whether she has understood the partner's contribution or not. Trivially, if the agent is not aware of her cognitive state, it is impossible to start reasoning on the relevance and acceptability of NewInfo, her response would be an unconscious reaction rather than a rational answer. However, it is not necessary for the agent to believe that she has understood the contribution correctly or fully: she can express her need to obtain clarification or confirmation of what the partner had said (signal need for confirmation or clarification). If the agent believes that she has not really understood the contribution, it is of course impossible to estimate the relevance of the partner's contribution, but in this case it is still possible to accept the contribution as a valid turn, and signal her problems in understanding explicitly or implicitly (signal non-understanding).

The cases where the agent accepts the partner's contribution as a valid turn are different from situations where the agent has not accepted it. In the former case the agent also takes some responsibility for constructing the shared knowledge with the presented NewInfo, while in the latter case, the agent simply signals problems and tacitly leaves the responsibility for the partner to clear up any misunderstandings and irrelevant NewInfo with the partner. Note that if the agent does not accept the relevant NewInfo as being in accordance with her evocative intentions, the situation is a conflict, and requires further negotiations to resolve the problem. Conflict situations thus presuppose understanding and relevance of the contributions, but the agents have contradicting goals and they need to negotiate as to which, if either, will be the joint purpose.

The actual realisation of the agent's response depends on the agent's contextual reasoning as she evaluates NewInfo with respect to her own intentions and expectations, the goals of the dialogue, and thematic relatedness to the dialogue topic. For instance, the choice between the questions *why* and *when* as alternative continuations of the dialogue depends on whether the agent wants to know the partner's motivation, or whether she considers the partner's statement legitimate and wants to know more.

If the agent has not been able to understand NewInfo, she can also try strategies so as not to expose her ignorance to the partner: she can for instance take a risk and continue the dialogue as if having understood and accepted the partner's contribution, or she can say something very general and invite the partner to continue, thus eliciting more information from the partner wishing that it could also help her to interpret the previous NewInfo.

It must be emphasised that the agent may have misunderstood the intended meaning of the partner's contribution anyway, so from the partner's point of view, the agent has not understood the contribution at all. In these situations, however, the agent's reaction usually signals non-understanding implicitly, which then allows the partner to remedy the problem. This kind of implicit correction while smoothly continuing the dialogue is common in human conversations: possible misunderstandings and problematic references often get unnoticed as part of the general clarification strategies and information flow, cf. Heritage (1984). In fact, it can even be argued that communication is an attempt to remedy non-understandings through implicit feedback: as the dialogue unfolds, possible misunderstandings will usually become apparent and can be corrected through rephrasing and providing further information.

On the other hand, if the agent is aware that she has not understood the partner's contribution but still carries on, the question arises whether this kind of behaviour can be considered cooperative in the sense of Ideal Cooperation. Several ethical considerations are put forward concerning false behaviour, with severe punishment as a consequence of being caught lying. From the rationality point of view, false behaviour is not rational either since in the course of time, and if the agents have many encounters, it prevents the agent from achieving the goal efficiently. As discussed in game theory (Axelrod, 1984; Gymtrasiewicz & Durfee, 2000), in prisoner's dilemma situations where the agents can win if they both cooperate although not as much as if the other faked, and lose if they both fake but not as much as if the other did not fake, it is more beneficial to cooperate in the long run (the so called tit-for-tat strategy).

The agent may also notice a communicative attempt by the partner but not be able to perceive it appropriately. If the agent accepts this as a valid contribution, she will report the failing conditions to the partner and invite the partner to repeat the contribution; otherwise she will continue with her own goals and ignore the partner. Of course, the partner's communicative attempt may also simply pass the agent unnoticed, so the agent continues the dialogue as if nothing has happened. In both cases, however, the agent signals problems in perception, and the partner may want to re-try communication later. The context and the agent's non-verbal behaviour usually reveal whether the communication attempt is unnoticed or pretended to be unnoticed, and thus control the communication.

Finally, the agent can also reject the partner's contribution altogether. The agent usually rejects NewInfo if it is irrelevant or not understood, but can also reject it if it is not in accordance with the agent's own goals (conflict situation). It is worth noting that an incomprehensible or irrelevant response is not rejected automatically: a cooperative agent may choose to accept the response as a valid contribution and explore the reasons for its incomprehensibility or irrelevance by taking the initiative in order to clarify the presented NewInfo. On the other hand, rejection may occur even though the partner's contribution is relevant and comprehensible: there may be a real conflict between the participants' goals, and rejection stems from strategic reasons (cf. Goffman's's Strategic Interaction (Goffman, 1970) and cognitive and ethical considerations in Ideal Cooperation).

4.6   Information Management and Rational Agents

Let us now study more closely how rational agents manage the information flow in dialogue situations using considerations of rationality and communicative principles. Cooperation emerges when the agents fulfil the expectations of rational motivated action and the interpretation of their contributions can be based on the simple assumption that the contribution has 'meaning', i.e. NewInfo, in the context. If this appears to be irrelevant, the hearer can conclude that some of the basic communicative functions have failed (the partner did not understand, perceive or she was not in contact).

The agent has the initiative if she has initiated the conversation or introduced a new topic. By definition, a rational agent can take the initiative in the dialogue, and the partner is obliged to evaluate the agent's goals with respect to her own intentions, and as long as the requested initiative and assistance is in accordance with the principles of Ideal Cooperation. The agent's right for taking the initiative is restricted by certain roles and activities (e.g. teachers vs. students; giving a speech vs. dinner discussions, chatting vs. formal meetings), and there are also different culturally determined obligations that guide the agent's initiatives in socially acceptable ways.

By taking the initiative the agent "owns" the topic, and has the owner's right to determine whether the topic is discussed satisfactorily. Of course the agent cannot force the partner to discuss a topic, but once the partner accepts the agent's initiative, she also accepts to cooperate with the agent and commit herself at least to the basic communicative goals: to evaluate the NewInfo and report the result back to the agent. Obligations concerning the start and closing of the dialogue are related to culture specific politeness codes and the speakers' roles, in the activity they are engaged in, and can contain uncomfortable aspects of imposing obligations to the partner. Elaborated opening and closing sequences have been developed so as to reach a shared agreement about opening and closing of a topic without offending or losing one or the other's "face" (Garfinkel, 1967; Goffman, 1969).

Jokinen (1996b,c) defines the following obligations which the agents must fulfil in order for a dialogue continuation to be coherent. These obligations are applied when the agent determines the view-point for presenting the NewInfo, and thus they make use of contextual factors such as initiatives and the agent's goals.

These obligations are derived from the agent's competence to express the appropriate message adequately. "Current CC" refers to the CC of the partner's last contribution.

1. If the agent cannot evaluate NewInfo, the agent can shift CC to NewInfo in order to clarify the NewInfo.
2. If the agent has the initiative and the partner has produced an utterance with a thematically related NewInfo, the agent is obliged to accept the utterance, and can either shift the CC to a new one or maintain the current CC (continue with the same topic), depending on whether she considers the current CC fully specified or not (topic is closed or not), respectively. Usually there is an asymmetry between the system agent and the user here: the user can shift to a brand new topic, while the system is only allowed to introduce topics related thematically to the task at hand. The agent may also give positive feedback about the acceptance, depending on the politeness codes and cultural obligations.
3. If the agent has the initiative and the partner has produced an utterance with a thematically unrelated NewInfo, the agent can reject the shift and maintain the current CC (i.e. not accept the topic shift but continue with the current topic). Again, other obligations may require the agent to provide feedback about the reasons for rejection.
4. If the agent has no initiative nor own goals to pursue, the agent can accept the CC introduced by the partner, and continue the dialogue with the CC.
5. If the agent has no initiative but has her own goals to pursue, the agent can either continue with the partner's current CC (accept the topic) or shift the CC (revert to a previous topic or introduces a new topic), depending on whether the partner has shifted CC to a thematically related one or not, respectively.

By default, the agents continue with the same topic until it is mutually closed, i.e. accepted by the agents as being sufficiently clear so that the underlying communicative goal can be agreed as having been achieved, or the topic is rejected explicitly. Topic shifts occur if the presented NewInfo needs clarification (i.e. the agent finds the evoked response only partially related to the context or does not understand it), if information concerning the topic has been exhausted (there is nothing more to be said about the topic), or if the agent's interest is directed to something else. On the other hand, if the agents are interrupted by something, like a telephone call or somebody rushing in, or their attention is distracted by something that happens in the environment, they can shift the topic to the interruption, but once the interruption is over, the dialogue is usually resumed at the original topic.

A topic shift to NewInfo is commonly termed as a sub-dialogue: the main topic of the dialogue is stacked, and the NewInfo is moved into the centre of attention. Sub- dialogues are often treated as some kind of interruption in the main course of the dialogue, since the original topic is resumed once the new topic is closed.

We regard topic shifts to NewInfo as coherent continuations of the dialogue; indeed, they are often the driving force in conversations as the speakers inspire each other to bring forward new pieces of information to be shared. The resumption of the previous topic is a kind of side-effect of the agent evaluating the partner's contribution in the dialogue context and determining the appropriate joint purpose on the basis of communicative principles. For example, in the dialogue in Table 4.6, the sub-dialogue U7-W7 concerns the user's clarification of the NewInfo presented by the system in W6 (type of insurance service). Once the system has cleared this up in W7, the user can evaluate the NewInfo of W6 again (the evaluation of NewInfo in W7 was successful as it provided an answer to the question in U7, so the only ungrounded piece of information is NewInfo of W6). The user reports the result of the evaluation by providing the requested answer. Resumption of the previous topic (information on car insurance) is thus a consequence of the changed context in which the partner's contributions are evaluated and a response planned.

Being rational agents, the speakers do not shift topics arbitrarily. Topic shifting follows the coherence requirements of the dialogue, and these are rooted in the principles of ideal cooperative communication. Communicative responsiveness requires the agent to report the result of the evaluation of the partner's contribution, and consequently, the reporting is related to what the partner said, especially to the new information being conveyed. The communicative competence of the agent is shown in the way in which she presents the NewInfo of her contribution from a coherent view-point, i.e. selects the Central Concept of her message according to the dialogue context.

Table 4.6 Example dialogue

Agent	Utterance	TOP	NewInfo
U6:	Have you any information on car insurance?		Exist (information = car insurance)
W6:	What type of insurance service do you want?	Exist (information = car Insurance)	Insurance Sub-type = ?
U7:	What type of services there are available?	Insurance Sub-type = ?	List Of Sub-types = ?
W7:	You can choose one of the following categories: Insurance administration services, Insurance agents, Insurance brokers, Insurance companies, Insurance consultants	Insurance Sub-type = ?	List Of Sub-types = {ins.admin, ins.ag, ...}
U8:	Insurance agents please	Exist (information = car Insurance)	Insurance Sub-type = insurance agents

A coherent dialogue is easy to follow and clear in its goals. The coherence and clarity should not be confused with brevity and straightforwardness of interaction. While the latter are usually related to the efficiency of the dialogue (and thus to usability of the dialogue system that produces such interaction), coherence describes how successful the dialogue is in the exchange of new information between the participants. In other words, in a good dialogue the exchange of NewInfo is presented and grounded in a fair amount of time in a manner that smoothens out possible problems according to principles of cooperation and consideration towards the partner. We can assume that good dialogues are efficient in building the shared context between the participants efficiently. Efficiency can be measured in terms of NewInfo which is introduced and grounded in the dialogue.

As emphasised earlier, in a mobile and multimodal context, the agent may have well-specified tasks to perform, but the dynamic environment may require rapid decisions that change the tasks and goals. The agent must thus rely on communicative knowledge and the coherence of dialogue, defined with respect to the world knowledge, in order to successfully achieve its goals.

We distinguish between Central Concept and NewInfo on the basis of their information status in the dialogue context. The Central Concept encodes the topic that the dialogue is about, while NewInfo encodes the novel information about the topic which the agents exchange in the dialogue. On a surface level, Central Concept can be implicit in the contribution or referred to by a pronoun, while NewInfo must always be explicit in the contribution and it cannot be pronominalised.

The principles that govern the acceptance, maintenance, shifting and rejection of CCs are derived from the speakers' rational agency. Coherent shifts of CC in a dialogue are based on domain relations which provide expectations about the likely next CCs both in analysis and generation. If there are multiple candidates for a next CC, a heuristic ranking of discourse referents is used to select one.

NewInfo is important, because the planning of a system response starts from it: NewInfo is the main information to be communicated to the user. Central Concept provides the view-point from which the NewInfo is presented to the partner, and it is important in relating the contribution to the previous discourse. The CC takes care of the local coherence in the dialogue: it provides the coherent view-point for presenting NewInfo.

Coherence is understood as consisting of inferable links between ideas, objects and events referred to in consecutive contributions. The linking of central entities (events or objects participating in the events) is based on domain or world model relations: for example, when talking about a hiring event, possible coherent next topics include the hire object, the hire location and the reason for the hiring. We assume that the driving force is the communicative competence of the partners, as they analyse and evaluate contributions according to the communicative obligations in a dialogue context. Dialogue coherence in terms of identifiable links between discourse referents is then an expression of the fact that the higher level communicative obligations have been appropriately addressed.

5

Dialogue Systems

This chapter gives examples of how CDM has been implemented in a working dialogue manager. We start with the general requirements for dialogue systems, drawn from the theory of communication as cooperative rational action. These also serve as the main evaluation criteria for flexible, conversationally adequate systems.

5.1 Desiderata for Dialogue Agents

The desiderata for a communicatively competent, natural language-based interactive computer agent include the following (Jokinen, 2004):

- Physical feasibility of the interface.
- Efficiency of reasoning components.
- Natural language robustness.
- Conversational adequacy.

In system evaluation, these aspects can be used as the main evaluation criteria, on the basis of which the detailed quality features can be defined (cf. Möller, 2002, 2005).

5.1.1 Interface

Communication has fundamental constraints on physical and biological levels (see Section 3.1). In the interaction between a user and a computer agent, these constraints concern the user's perceptual, cognitive, and physical constraints, and are related to her short-term memory span, cognitive load, eye-sight and hearing, capability to use hands to type or move the mouse, etc. They form the first important requirements for flexible, user-friendly conversational computer agents: interaction should be physically easy and feasible. Usually these factors are studied in the ergonomics of the interface, and the findings are integrated in the design of the actual interface device and in the design practises of usability (the system is easy to use, easy to learn and easy to talk to) and transparency (the system's capabilities and limitations should be evident to the user from experience). However, as argued in Section 1.3.4., the ultimate human factor in interface design is the system's conversational ability. Different input and output modalities also contribute to the system's robustness, and one of the supporting arguments for multimodal systems is that they allow for natural interaction (the possibility to use all communicative modalities) and freedom (the user can choose which modality to use in a particular situation). From the Design-for-all point of view physical feasibility is important not only because it supports accessibility for users with special needs, but also because the interfaces are commonly more usable if the general principles of accessibility are taken into account. Given situated computer agents, multimodality is a default rather than an extra feature for them.

5.1.2 Efficiency

The speed of the agent is of course important: real-time processing is demanded of the interactive systems if they are to be usable and useful. Nielsen (1993) argues that the tolerable waiting time for a computer response is about 10 seconds, while Nah (2004) noticed that for web users the perceived waiting time appears to be more important than true waiting time. Consequently, techniques that decrease the amount of time the user perceives as having waited, are pertinent in system development: for example, feedback prolongs the web user's tolerable waiting time, and is thus useful. Other characteristics such as modality and complexity of the interface can also influence the perceived waiting time. Studies also show that the users can forgive slow reaction times if the system as such is interesting and novel (Jokinen and Hurtig 2006).

When considering spoken dialogue systems, the users usually acknowledge the naturalness of system responses, but still regard long response times as one of the most negative features in their system evaluations, and one of the main factors that distinguishes computer dialogues from human-human conversations. If the partner fails to reply in a reasonably short time, the agent normally starts to think that there is something wrong with the basic enablements for communication (contact has been lost, the partner couldn't perceive the act, or didn't understand it) or, if these seem to be fine, that the partner fails to observe the obligation of responsiveness and consequently, causes the requirement of trust of Ideal Cooperation on the agent's side to be broken (the computer is perceived as rude and impolite). Long response times can cause unnecessary turn taking and end with the user's frustration which are undesirable features when interacting with service agents and practical applications. In the ubicom environment it is even more important to allow for real-time operation and maintain the system's reliability and user-friendliness: since the computer agent must be able to participate in the interaction as an independent and reliable partner.

In addition to faster hardware and communication technology, more efficient algorithms and various processing heuristics are also needed: predicting the most likely next actions, providing fast reactions for frequently occurring discourse phenomena and allowing for more processing for those that occur rarely, etc. Another solution is to employ methods of narration and interactive storytelling in order to design interaction patterns where the necessary input processing is interleaved with presentations to the user (see e.g. Laurel, 1993). This solution accords with the ACA view of the interaction where the user is involved in two activities with a computer agent (Section 3.2.2): the communicative and non-communicative activity, and the agent tries to address both of them by maintaining the conversation while simultaneously executing relevant task-related commands.

5.1.3 Natural Language Robustness

Linguistic sophistication has often been questioned as being esoteric and rather peripheral in interface design. The number of words used in human-computer interactions is often limited, and interaction can be designed so as to consist of a set of short commands and selected answers. In such cases natural language enhancement, e.g. ellipsis, does not necessarily improve the system's usability, and in some cases such enhancements can even obscure the task, which would be completed most efficiently by direct manipulation interfaces or with the so called tangible interfaces (Milekic, 2002).

It is clear that verbal interaction is not the optimal mode for executing tasks such as design and graphics, text-editing, parameter selections for hotel bookings etc., and other modalities are better suited for their interactive achievement. However, negative conclusions about the use of natural language are seldom due to the use of language as such, but to a simple underlying task that does not lend itself to more complicated verbal interactions. The emphasis on the dialogue system's efficiency has directed applications towards well-structured tasks where the use of natural language can be reduced to a simplified command language that does not allow the power and positive characteristics of the language use to evolve. When considering more elaborate tasks which require abstract processing of different conceptual alternatives, the computer agent's communicative capability as well as the quality of language processing components become essential. In these cases, intelligent communication increases the system's usability by reducing such unintended problem solving situations as "what would be the best way to put the questions so that the system would understand me?" The use of natural language thus makes the system affordable, see Chapter 6.

5.1.4 Conversational Adequacy

Conversational adequacy manifests itself mainly in the contributions that the computer agent produces in order to clear up vagueness, confusion, or lack of understanding which occur in user contributions. When interaction proceeds smoothly, there is not much difference between the three types of interaction models that the agent may realise (reactive, interactive, and communicative models, Section 2.1), except maybe in processing speed and implementation effort, but when there is a need to negotiate correct meanings for references, clarify vague expressions and produce cooperative responses that would help partners to achieve their goals, the communicating computer agent should manifest its superiority to conduct intelligent communication.

For instance, rather than viewing a clarification request as a signal of failure, the agent can model it as a normal, cooperative reaction to the information presented earlier. The dialogue resembles negotiation which continues until the dialogue partners have achieved their goals and they are satisfied with the result. The information exchanged must be relevant at a given state of a dialogue, but it need not be exhaustive as long as it is appropriate with respect to the dialogue situation.

Given that the agent is capable of sensing its environment and interpreting the signals it receives from the context in a meaningful way, it may also show its understanding of the users by various non-verbal signals and by demonstrating its cooperation through adapting its actions according to the perceived user characteristics, skill levels, and the general conversational atmosphere. All this brings conversational adequacy into the focus of the challenges that advanced dialogue interfaces encounter (cf. arguments already in Jokinen, 2000): the intelligent computer agent should learn through interaction, update its knowledge bases, and adapt its output to new situations appropriately.

5.2   Technical Aspects in CDM

Agent-based architectures provide suitable ways to implement ACA-type dialogue management: the basic reference architecture can be extended and modified with other modules that learn to react to the appropriate system states as required by the task. The architecture thus exhibits flexibility which makes the system processes adaptable to various communicative situations, while the dialogue model functions by specifying the overall goal for the software agents to act in a way that follows certain robust ways of producing desired reactions.

5.2.1   Distributed Dialogue Management

Kerminen and Jokinen (2003) propose a distributed dialogue management model, where dialogue decisions are distributed among the system's components. This view accords with the other agent architectures (OAA, Communicator, TRIPS, etc., see Section 2.3.1). The distributed dialogue management means that the decisions about how to react to a particular input are distributed over the whole system, and are not handled by a single monolithic DM. Moreover, input and output processing are interleaved with dialogue management, and task management is separate from dialogue management.

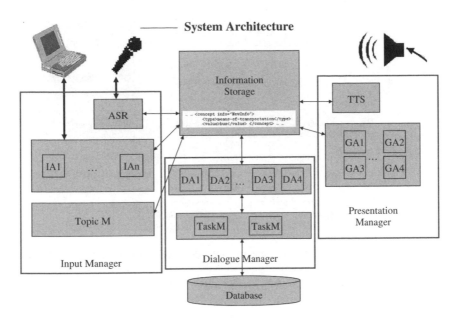

Figure 5.1 Dialogue system developed in Interact

Figure 5.1. depicts the architecture used in the Interact system (Jokinen et al., 2002). The architecture follows the logical flow of information (see Section 1.2.3), and consists of the interpretation of the input (Input Manager), dialogue management (Dialogue Manager) and response planning (Presentation Manager). Moreover, the system maintains a dialogue context (Information Storage), and takes care of several component interfaces (database search, interfaces to other components).

The system is built on top of the Jaspis framework developed by Turunen et al. (2005). Jaspis is a general blackboard-type architecture for speech-based applications, and the key characteristics of the architecture are the shared information storage and the concept of managers, agents and evaluators (the terminology is somewhat unfortunate since the agent refers to the actions available for a manager, while the manager corresponds to what is commonly called the agent in agent-architectures). The major components of the system thus consists of agents that are specialised for certain tasks like input analysis or dialogue actions, and managers which group the agents and coordinate their suitability to a particular dialogue situation by the help of evaluators. Evaluators score the agents, and the agent with the highest score is selected as the action of the manager. (see Figure 5.5 in Section 5.2.3).

Dialogues are modelled as series of dialogue states. Each move provides new information to the hearer about a particular topic in the context, and it results in a new dialogue state as shown in Figure 5.2. The features of the state include

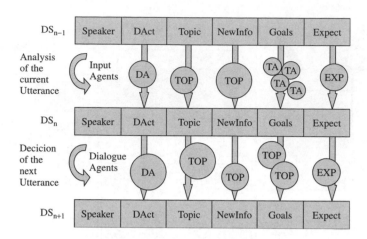

Figure 5.2 Dialogue states and dialogue agents. DS_n describes the state after the user utterance n, while DS_{n-1} and DS_{n+1} describe the state after the system utterances $n-1$ and $n+1$. The circles represent dialogue agents producing various information

the speaker, which links the state to possible speaker characteristics, the dialogue act (DAct), which describes the act that the speaker performs by the particular utterance, the topic (Top) and the new information (NewInfo), which denote the semantic content of the utterance, the unfilled task goals (Goals), which keep track of the activity related information still necessary to fulfil the underlying task (approximates a plan), and the expectations (Expect), which are related to communicative obligations and used to constrain possible interpretations of the next act. The system's internal states are thus reduced to a combination of these categories, all of which form an independent source of information for the system to decide on the next move.

The dialogue manager and task manager communicate with each other via a particular task-form which has as its parameters the concepts important for the task manager to fetch information from the database. If the form is filled in so that a database query can be performed, the task manager returns the form with all the appropriate parameters filled in, and thus lets the dialogue manager decide on the status of the parameters and their values with respect to the dialogue situation. The application domain is public transportation and bus timetables.

Figure 5.3 shows how the system works. Dialogue level input analysis is handled in the Communication Manager, task related information in the Task Manager, output generation in the Presentation Manager and decisions of what to say next in the Dialogue Manager (see more in Jokinen et al. 2002; Kerminen & Jokinen, 2003). As already argued, distributed dialogue management supports the rich functionality of dialogue systems and adaptive interaction as components and agents are in principle easy to add. For complex tasks, there may be various possible task decompositions, and the opportunity to design different software agents for their implementation facilitates the system design.

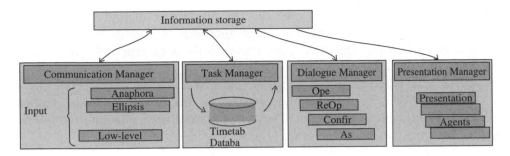

Figure 5.3 Dialogue management in Interact

There are also some questions for which the dialogue system developer must also provide answers. For instance, knowledge representation is an important issue and representations that are logically solid and linguistically expressive should of course be preferable, cf. 2.3.2. Another issue to deal with is the decomposition of the task. As the complexity of the task increases, there are many ways to conceptually divide it and provide compositions that are possible implementations for the system. It may be possible to establish a small number of basic software agents and agent types that can be used as building blocks for building "basic" systems, and which can then also be used to build systems with more advanced possibilities, which respond adequately to various design requirements. Adaptation and learning aspects are important in the system design as well, and they will be discussed more later: adaptation on the architectural level will be investigated in Section 5.2.3 in connection with adaptive agent selection, while adaptation to the user strategies and skill levels will be studied in Section 6.1.

5.2.2 Generation from NewInfo

The driving force behind the dialogue is to exchange information on a particular Topic by providing new information, NewInfo, to the partner. It should be noticed that the building of shared context is not a monotonous unification of compatible new information but can, in fact be destructive in that the system may correct previous erroneous information. As an example of how the NewInfo approach works in dialogue modelling we study how NewInfo and Topic (CC) are used in response generation, see more detailed description in Jokinen & Wilcock (2003). The starting point is the information exchanged in the dialogue, and the basic unit for this is the notion of NewInfo. As argued in Jokinen (1996a, c) and Jokinen et al. (1998), the NewInfo is the only information that is required to be included in the utterance: unless the context does not disambiguate NewInfo, the utterance needs to be wrapped with the necessary explaining concepts. The construction of a shared dialogue context takes place by accommodating NewInfo into the agent's own understanding of the dialogue situation and providing implicit or explicit feedback of the success of this accommodation to the partner. Although higher-level expectations of the goal and the appropriate dialogue strategies are needed to guide the reasoning, the basic approach is bottom-up: through the identification of the NewInfo and its unification with the current dialogue situation, interaction is managed and constructed locally.

Jokinen and Wilcock (2003) illustrate the effective use of the concepts of NewInfo and Topic for the planning process that lies at the border of dialogue processing and tactical generation. They consider response planning for the two related but syntactically different questions:

(1) Which bus goes to Malmi?
(2) How do I get to Malmi?

In the case of (1), Topic contains the concepts "means-of-transportation" and "Malmi" (according to the question method discussed in Section 4.2, it can be paraphrased as "talking about buses to Malmi, which one takes me there?"), while NewInfo concerns bus numbers. In the case of (2), however, NewInfo can be related either to a literal question about public transportation to Malmi, or to an indirect request for buses that go to Malmi. In the literal case, the Topic of the exchange is regarded as "Malmi", (paraphrased as "talking about Malmi, how do I get there?") and the NewInfo as "means-of-transportation", whereas in the indirect case the Topic contains the two concepts "means-of-transportation" and "Malmi" (it can be paraphrased as "talking about means-of-transportation to Malmi, what is the type that takes me there?"), and the NewInfo comprises "type=bus" and "busnumber". The difference in the information structure between the indirect request and the request in case (1) reflects the difference in the speakers' presuppositions: the request in case (1) assumes that there is a bus that goes to Malmi, while no such assumptions is made in the indirect case.

However, in all cases the information that the dialogue manager gets from the task manager is the same: there exists a means of transportation to Malmi, i.e. a bus and its number is 74. The response generation for answers follows the information structure of the questions, so that the new information of the answer accords with what is requested as new information. The simplest realisation is to realise NewInfo only and leave the topic concepts out. In the case (1) this is actually possible, and the response becomes:

(3) Number 74.

In the other cases, however, the situation is somewhat more complicated. In the literal case the requested NewInfo, i.e. "means of transportation", gets privilege, but since the NewInfo also comprises the concepts "type=bus" and "busnumber", they represent extra information with respect to the original question and they may or may not be realised in the response as well. In service systems, the assumption is that the system behaves in a cooperative manner and thus the inclusion of the extra helpful information in the response can be seen as a sign of cooperation and consideration on the system's side. However, the extra information need not always be realised, i.e. the computer agent need not tell the user all the information it knows. This depends on the relative status of the information in the dialogue context (it may be irrelevant, uninteresting, secret, etc.), the user's access rights, and the agent's general cooperation settings. Concerning the indirect request, however, both "type=bus" and "busnumber" are considered part of the relevant new information to be given to the user, the first addressing the user's presupposition about the existence of "type=bus", i.e. bus service to Malmi as opposed to local train or metro, and the second about the factual information about the bus service. The difference in the information structure is reflected in the surface realisation of the answers alternatives (4) and (5), corresponding to the literal and indirect questions, respectively:

(4) By bus - number 74.
(5) Bus 74 goes there.

5.2.3 Adaptive Agent Selection

Recently, reinforcement learning (Barto and Sutton 1993) has been used in learning dialogue strategies and dialogue management (see references in Section 2.1.4). Reinforcement learning is a semi-supervised learning algorithm, and the framework can be depicted as in Figure 5.4. The agent performs an action "a", the action changes the environment into a new state "s", and the agent receives a reward "r" (together with the observations of the new state "s"). The task is to find a policy that maximizes the agent's reward in the environment. The formal definition of the RL framework consists of a discrete and finite set of states, a discrete set of actions that the agent can take to go from one state to another, and a scalar valued reward function which describes the reward the agent gets when

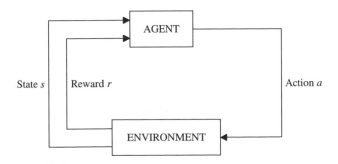

Figure 5.4   The reinforcement learning framework

it performs an action in a certain state. The policy is a function from states to actions, and it describes the agent's behaviour in a given state.

Q-learning (Watkins, 1992) is a technique to estimate the values of state-action pairs (Q-values), $Q(s,a)$, by learning which action is optimal for each state. It does not need an explicit model of the agent's environment and can be used on-line. The q-values are calculated as the expected discounted sum of future payoffs when an action "a" is taken from state "s" and an optimal policy is followed thereafter. Once the Q-values have been learned, the optimal action from any state is the one with the highest Q-value.

Kerminen and Jokinen (2003) pointed out how agent selection in the Interact architecture could be interpreted as action selection of autonomous agents, and how the selection could be implemented using reinforcement learning algorithm. As mentioned, the agents in the Interact system are not autonomous and intelligent software agents but rather actions available for the manager. Each agent (action) is scored by the evaluators according to the agent's applicability within a given dialogue situation, and the one that receives the highest score is selected.

Figure 5.5 depicts the heuristic agent selection where the CanHandleEvaluator selects the agent that can best handle a particular dialogue situation. The agents represent different dialogue acts, and they are scored on the basis of how well they suit to the dialogue situation (if it is an opening or closing situation, confirmation or question asking, etc.).

Figure 5.6, on the other hand, shows agent selection based on reinforcement learning. Now the evaluator is replaced by a QEstimate evaluator which chooses the agent that has proven to be most rewarding so far, according to the Q-learning algorithm.

The evaluator is based on a table of real values, indexed by dialogue states, and updated after each dialogue. Adaptivity of the dialogue management can thus appear on the level of system architecture with the distributed dialogue components, as well as on the level of the components' internal functioning based on the reinforcement learning algorithm in the QEstimate evaluator. More experiments

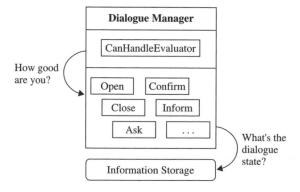

Figure 5.5   Heuristic agent selection in the Interact-system. Kerminen & Jokinen (2003)

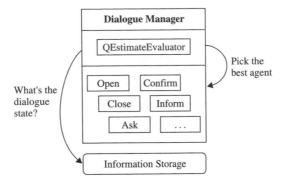

Figure 5.6 Reinforcement-based learning in the Interact-system. Kerminen & Jokinen (2003)

on the reinforcement learning algorithm applied to the online learning of user strategies will be described in Section 6.1.2.

5.2.4 Multimodal Route Instructions

As already discussed, full-scale communication presupposes that communication can take place through different modalities. In this section, we will briefly discuss multimodality issues related to the MUMS system which focused on route guidance and navigation with the help of speech and map-gesture interaction (Hurtig and Jokinen 2005; 2006). The system is a continuation of the Interact-system, and provides the user with local transportation information.

The MUMS system is accessed from a standard PDA device, and the users can interact with the system using natural speech and map gestures. The PDAs have GPRS connection to the server where all the processing of the information takes place, except for the light weight speech synthesis in the client devices. The system also utilizes GPS information to locate the user, thus simplifying the dialogue. Figure 5.7, taken from the MUMS website (http://www.ling.helsinki. fi/~thurtig/MUMS/index_en.html), depicts the MUMS system.

Figure 5.8, also from the website, shows how MUMS interaction takes place: the user requests information using speech and a pointing questure and the system responds by speech and drawing the route on the map. The system responses are also stored in a written format so the user can choose to read them if the environment is too noisy to listen to the responses. The touch-screen map interprets all tactile input as locations: tap denotes a particular location and a circle is a set of possible locations in that area. The map can also be freely scrolled and zoomed in real time.

The system's dialogue strategies concern confirmation and initiative strategies whereby the system confirms the intended referents of the user's requests, and guides the user through the route, respectively. In the latter case, the users may be allowed to interrupt and enter into "explorative" mode, if they want to explore the environment more (Wasinger et al., 2003).

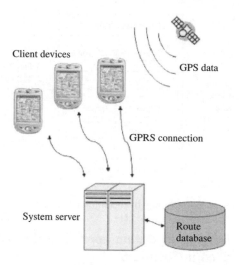

Figure 5.7   The MUMS system

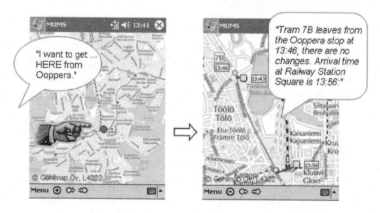

Figure 5.8   MUMS interaction: on the left the user's request and on the right the system response

The interpretation of the inputs is done in three phases whereby the time-stamped inputs are combined to form an interpretation of the composite input. The fusion consists of the production of all possible combinations of speech and tactile inputs, their weighting with respect to proximity and type, and finally the selection of the best candidate. If none of the candidates fits in the current dialogue context, the user is asked to repeat the question. A more detailed description of the fusion can be found in Hurtig and Jokinen (2005).

The MUMS system represents a preliminary version and experiments with a system that would allow natural and flexible interaction "anywhere anytime". However, in order to enable natural and flexible communication in mobile and ubiquitous situations, two larger issues need to be addressed: the basic enablements of communication and the grounding of situational information. Concerning the basic enablements, the design must take into account the fact that the user's full attention is not always directed towards the device and the service that the device provides, but to the primary task that the user is engaged in, such as driving a car, talking to people, taking part in a meeting etc. The computer agents thus need to be sensitive to the user's current situation and their interaction capability be tailored so that they support the user's needs: they should be easily available when the users want them and when they can use them. This requires awareness of the context and understanding of the user's goals, focus of attention and emotional state. Moreover, analysis of the CPU requirements, contact, perception and understanding needs to performed so as to understand the necessary conditions and prerequisites for a successful exchange of information in a particular interactive situation. Given that the user's environment contains several intelligent applications and devices, coordination of their communication also becomes a relevant and urgent research topic. In human-human communication, much of the current research deals with issues that are related to CPU requirements, feedback giving processes, and multiparty communication.

The other main question deals with the grounding of information. Especially in robot communication and location-based applications dealing with navigation and way-finding the user's needs are related to the immediate and changing environment, and one of the problems is the frequent reference to spatial information. The speakers refer to their physical environment, and the spatial and visual environment directs their thinking. Usually the environment and the user's possible location are taken into account as a fixed context for the interaction design, but the development of sensor technology and communication technology also allows for more dynamic handling of spatial referents: intelligent agents can respond to grounding issues using their observations from the real environment. However, the speakers' mental model of the environment is not necessarily a one-to-one mapping with the geographical data but rather a view of the environment filtered through the speakers' own understanding and attitudes, see e.g. Tversky, (2000). References can thus be inaccurate and vague, and the speakers' dialogues are prone to misconceptions and misunderstandings. However, humans usually handle the situations by providing more information or asking further relevant questions so as to clarify and agree on the correct reference, see e.g. Skantze (2005) for human error handling strategies in the context of pedestrian navigation and guidance. In AI, grounding has been regarded as one of the most important problems, see Harnard (1990) for a discussion. See also Clark & Wilkes-Gibb (1986) and Traum (1994) for grounding and construction of the shared context.

5.3 Summary

The Constructive Dialogue Model takes the view that interaction between humans should be regarded as rational interaction, and when designing intelligent interactive applications, the system's communicative capability should be taken into consideration. The desiderata for such interactive computer agents have been summarised as four general evaluation criteria referring to the different enablements for intelligent communication, with the criteria for conversational adequacy concerning the agent's communicative capability in particular.

As some concrete examples of how CDM has been implemented in interactive systems, the chapter also presented some solutions in building such systems: distributed dialogue management, generation from the new information, adaptation, and multimodality. In all cases, dialogues are managed depending on the current context, represented in the shared knowledge base, and responses are planned on the basis of ideal cooperation and the obligation of responsiveness. Cooperativeness, coherence and robustness are exhibited by the dialogue as a result of applying the general principles of communicative activity; and the observed relations between consecutive dialogue contributions, modelled with the help of structural rules, domain (in)dependent coherence relations or the goals and intentions of the speakers are tools for the analysis and evaluation of the contributions.

It should be noted that the example systems are designed to work as practical applications rather than to highlight certain theoretical principles, and thus they do not exhibit full-scale communication reasoning starting from the basic principles of cooperation and rational action. As an example of the latter, however, see Jokinen, (1996a, b), where the deliberation about appropriate responses is part of the system design, and Black et al. (1991) where abductive reasoning is extensively used to experiment with input interpretation and goal formulation.

Finally, it is important to notice that even though the CDM type deliberation can, and in practise usually is necessary to be divided into two separate tasks of interpretation and planning following the logical flow of interaction, conceptually the two tasks form one single process whereby the agent's reactions are basically just the end of the agent's evaluation of how the partner's contribution fits in the agent's own plans. The constructiveness of the dialogue model is thus found in the construction of the reaction through interpretation.

6

Constructive Information Technology

In this chapter the main themes of the book are discussed with respect to learning and adaptation. Extensions of CDM and "full-blown communication" are discussed related to human-human communication and human-computer interaction, and design principles for adaptive dialogue systems are reviewed from the point of view of communication as the human-factor in interaction design. Finally, aspects of the usability are re-analysed in terms of affordance that natural communication provides.

6.1 Learning and Adaptation

Human agents adapt themselves to the partner's style, words, knowledge level, language, etc. This is often called alignment (Pickering and Garrod, 2004); see also discussion in 3.1.3. concerning cooperation and construction of shared context. It shows that communication is a collaborative project where the partners try to make the understanding and exchange of information effective and easy by accommodating themselves to the current situation. Adaptation is thus essentially learning to cope in a particular situation, and consequently it can be modelled in a similar way to the cooperative interaction between an agent and the environment, i.e. adaptation is achieved by producing a series of changes that affect the agent's beliefs, knowledge, attitudes, and emotions. Of course, adaptation differs from the communication in that it is usually not planned whereas communication is characterised by certain goals that the agents intend to achieve. Thus it is not easy to measure adaptation: it is not possible to measure it with respect to whether a goal was achieved or not.

A common counter-argument against machine intelligence deals with learning systems (see e.g. McCarthy, 1979). On the other hand, it is not impossible to write a computer program that could learn over time and produce behaviour which is not readily predictable from the inputs. For instance, an ordinary internet search query on the web acts very much like a creative response: the same query retrieves

Constructive Dialogue Modelling   Kristiina Jokinen
© 2009 John Wiley & Sons, Ltd

a different set of answers for different users, even for the same user at different times, since the ranking of the hits is based on the previous search history.

Creativity thus appears to be a similar concept to intelligence in that it need not be formalised explicitly as part of the computer agent's knowledge but rather, it emerges through the user's interpretation of the computer agent's actions in a way that conforms to human standards of creativity. This of course presupposes that the participants trust each other to act in a rational way and in accordance with the principles of Ideal Cooperation.

As already argued in (Jokinen, 2005b), the problem in adaptive interfaces appears to be the notion of adaptivity itself: adaptation involves learning, learning involves interaction, and interaction changes the knowledge through which adaptation takes place.

One of the main issues in adaptation is when to adapt and what to adapt to. The customary situation now is that the user adapts to the system. The users learn to use a particular system, they use certain style in speaking and shortcuts in giving commands. The computer agent can also try to adapt to the user. It can model the increase in the user's knowledge and expertise levels, and thus allows for monitoring of the user's behaviour and, on the basis of these observations, tailor its actions and responses accordingly. Yet a third type of adaptation is available, namely system adaptation to different applications, devices and environmental parameters. This kind of system-level adaptation was already discussed in connection with agent-based architectures in Section 2.3.1. Within the ubiquitous computing paradigm, and combined with technology for sensors, multimodal input/output channels, location and context-aware systems, evolvable hardware, etc., it seems more plausible that a computer agent could be equipped with an ability to accommodate its behaviour to the needs of the user, and to adapt its actions in accordance with the requirements of the dynamically changing environment. Moreover, it has also been envisaged that the system should also be able to adapt to different software programmes and change itself to be a radio, a phone and a computer as is necessary for the user.

Adaptation leads to issues of usability where the main question is the desirability of adaptation. While adaptation is natural from the point of view of dialogue modelling and human communication, the opinions within the interface and system design communities are divided with respect to the benefits of adaptation.

The arguments are based on the view of the computer as a tool on one hand, and the computer as an agent on the other hand, and can be compared with those presented in the AI debates about the strong and weak AI, except that the focus now is on whether the usability of a system would increase if the system had some autonomy of its own that would allow it to react to certain user needs in the course of task completion. Those against this view emphasise that the user should be in control of adaptation and decide when the system should adapt or whether it should adapt at all, while those in favour argue that applications become so complex that it is impossible for a user to master the entire system, and adaptation by the system would be necessary. As the interfaces become conversational, the natural adaptation that is common in language communication is thus expected to manifest itself in human-computer interaction as well. As a result, the question of usability and desirability of adaptation can be formulated in terms of interactability, or, as we have termed it "conversational adequacy" (Section 5.1.4).

At this point, however, it is good to clarify the general assumption that the user adapts her actions when using computer applications. The users seem to have certain preconceptions of the tool and how to use it, and they do not venture into exploring the limits of the system, even though they are encouraged to do so. They seem to assume tacitly certain limits to the system, and their exploration of, and adaptation to wider system capabilities appear to be prevented by their knowledge of the computers and by their familiarity with IT technology in general. For instance, in our preliminary tests one subject requested a help-facility but never actually tested whether the system had such a functionality or not (the setup indeed contained help). Furthermore, some system functionalities were not commonly used although they were mentioned explicitly in the instructions. Thus, although the users are capable of adapting their behaviour to novel interfaces and interaction strategies, they usually act in a conservative way: they prefer interaction techniques that they are familiar with to those that would require more learning and adaptation. In order to get a task done quickly, this is of course the best strategy, but from the adaptation point of view, and especially considering interactions with computer agents that develop and change, this may not be an optimal situation. If the users' willingness to adopt new techniques depends on how useful and effective they think the new patterns are for the task at hand, these preconceptions may unnecessarily limit their views about what is possible. Unless the users have an explorative mind, they may only be acting out restrictive expectations rather than being inclined to adopt useful alternative ways to interact with a computer. Bad experience may thus counteract development in interaction technology and cut off potentially useful features as not usable. Although it may look easier to allow the user to adapt to the interface technology than to build models for the system adaptation, in the long run, it may actually prove more beneficial to bring intelligence and adaptation into the system itself, so as to support the user's learning process too.

A common problem in interactive systems is that they are static and usually provide only one type of dialogue strategy for the users to follow. Also the system's knowledge of the user is often restricted to a general level of user expertise: this undervalues the user's versatile competence which varies depending on the task at hand. The static nature of the systems is a problem especially for applications that are intended to be used in mobile and ubiquitous environments, by various users with different abilities and requirements. Below we will discuss adaptive user modelling and dialogue strategies from this view point.

6.1.1   Adaptive User Modelling

Jokinen (2005b) discusses adaptation in interactive applications, and divides it into static and dynamic adaptation. Static adaptation refers to options that the users have in making decisions on such interface aspects as colour or sound choices which depend on the user's personal preferences, and can be listed in personal profile files. Dynamic adaptation deals with the computer agent's learning ability concerning typical interaction strategies or user preferences. This may be realised in the classification of the users, e.g. on the basis of their explicitly expressed preferences or navigation choices, so as to provide better tailored answers to user queries or to produce educated recommendations to the user on a particular topic (films, books, cars, etc.). A realistic model of adaptation would also require that the computer agent's knowledge-base is updated dynamically, in other words, the agent's knowledge on which reasoning is based, is modelled as meaningful clusters that can change depending on the interaction with the user as well. This, of course, brings the problem back to learning and knowledge acquisition discussed above.

In the DUMAS project (Jokinen & Gambäck, 2004), one of the main goals was to explore dynamic adaptation and study ways in which the system can adapt to the user's skill levels. The application domain dealt with checking one's email over a mobile phone, and thus it was easy to assume that the user would not only have one-time interactions with the system, but would use it frequently so that adaptation could take place in the system over the time. Adaptation was taken care of by the User Model component which recorded the user's actions and estimated the user's competence levels, giving recommendations to other system components on the appropriate way of responding.

The specific model that concerns the user's competence level in DUMAS is called the Cooperativity Model. It calculates the explicitness level of the system utterances as well as the level of dialogue control exerted by the system in the interaction. The Dialogue Manager uses these values when deciding on the next action, and when producing an utterance plan to be realised as the next system response. It gives more guidance on how to use the system if the problems seem to originate from limited user expertise, and if the fault is in speech recognizer or in language understanding, the system assumes a more active role and takes more initiative.

The UM consists of an online and offline part which deal with the online interactions with the user and the offline operations on the data, respectively. Also the Cooperativity Model has online and offline components so as to react to specific situations at runtime, and to track long-term developments in the user's experience with the system. Both components use a number of automatically produced parameters that record the user's behaviour, but they calculate their recommendations slightly differently due to their different functions: the offline parameters change more slowly in order to round off coincidental fluctuation in the user's experience with the system, while the online module reacts rather quickly to the user's actions, so that the user's adaptation to the system functionality can be addressed immediately at runtime. The offline values are default values that are used when the user logs in the system, but they are modified by the online component following observations of the user's behaviour with the system. The default values thus represent the user's assumed skill level as it has been developed through her particular interactions with the system so far.

The Cooperativity Model monitors and records the user's actions specifically on each possible system act. This allows the system to adapt its help to the user's familiarity with individual acts, not just on the user's expertise in general. The parameters deal with such aspects as the number of help requests, timeouts, interruptions, repetitions of system acts, and repetition of the information content of the user request, and the online module also includes a dialogue problem detector that monitors speech recognizer errors. The parameters are weighted according to their frequency and recency, and to the number of sessions that the user has had with the system and to the number of sessions during which the particular system dialogue act has been invoked. The parameters and parameter weighting are in part based on those used by Walker *et al.* (2000) and Krahmer *et al.* (1999), while the initiative levels follow Smith & Hipp (1994). The details of the Cooperativity Model can be found in Jokinen & Kanto (2004) and in Jokinen (2006).

The evaluation of the Cooperativity Model showed that the system could observe and learn changes in the user's skills to interact with the system, although the time period for the evaluation (one week) was not long enough to consolidate the changes. Two important points in the development of adaptive systems came across, however. First, the system should act consistently, and its responses should be carefully designed so as to respect the user's need to understand the system's functioning and the overall goals of the adaptation. Second, the responses should pay attention to the type of information that is being exchanged. Often it is not only the amount of information that is important (e.g. an exhaustive help on how to use a particular command), but especially the kind of information (description of a certain command vs. its implementation in the system). For instance, in the early studies by Paris (1988) concerning explanation generation of how machines work, it was noticed that novice users benefited most from descriptions of the machines while experienced users preferred to know their functionality. Giving helpful information on some topics does not only require planning on how much detail will be given to the partner but from which view-point the information will be presented.

When considering the system's adaptation to the user, we must also pay attention to several other aspects that center around the user besides the user's skill level. First of all, the users have different attitudes and preferences that the system may need to adapt to. In fact, one of the most studied areas in adaptation is how to address the user's preferences and likings (cf. recommendation systems and e-commerce applications). Using the common method of collaborative filtering, the user's preferences can be compared with the preferences of a group of users with a similar background, and thus it is possible to augment the preferences of an individual user with the preferences that hold among people with similar background; it can be assumed that the background provides a reference framework within which attitudes and preferences of an individual are formed to match with those of the others. However, this kind of adaptation can easily become mechanical classification unless the classes are continuously updated and recommendations modified accordingly.

Another area of adaptation is the user's communicative strategies and dialogue management habits. The users may be active and take initiative or they may be passive and wait for topics to appear. The system could thus adapt its dialogue management so as to accord with the user's preferred style, assuming that it would be easier for the user to follow the system's recommendations and also to adopt suggestions if the system were similar to the user in the dialogue style. Ability to address the user's temperament and style is related to this kind of adaptation: users vary in how quickly and strongly they react to problems, and the system may try to learn these differences and tune its responses accordingly. This kind of adaptation will be discussed more in the next section.

6.1.2   Dialogue Strategies

Conversational strategies concern certain behavioural patterns that dialogue partners exploit in conversation. These patterns can be described as the agent's inclination to take initiatives or to follow-up the partner's initiatives, to confirm each piece of NewInfo or to move forward, or to present large chunks of information in one go or to give step-wise instructions, to provide background information or go straight to the point. These descriptive strategies can be pinned down in the use of particular act chains: the use of certain dialogue acts has a higher probability than the use of other acts.

Differences between agents' conversational strategies can be made clearer by visualising the dialogue act paths that the agents are most likely to use in their interactions. For instance, Jokinen *et al.* (2002) describe such experiments as part of the design of their planned email application. Data was collected via Wizard-of-Oz setup with a scenario-based group interaction (see Kanto *et al.*, 2003) and users were asked to check their mailboxes with the experimental system that incorporates "normal email client functionality". During the trial period of one week, the users were requested to phone the system at least twice a day, so as to create enough email activity and to allow for development of the topic.

Figures 6.1 and 6.2 visualise differences in possible state transitions learnt from the data. The reinforcement learning algorithm (see Section 5.2.3) has been applied to online learning of the user's actions and action sequences. The transition possibilities in the action state space are represented as a tree and the Q-values are associated with each transition; the optimal path through the state space is thus marked by the transitions with the highest Q-values. The reward function has simply been the transition probability moving from a particular state to another state in the corpus.

Figure 6.1 shows possible dialogue paths averaged over all users while Figure 6.2 depicts dialogue paths for one particular partner. Although the transition paths are based on a small corpus, they suggest that there is a clear difference between the two cases: the general pattern (Figure 6.1) seems to be to read the message first and then repeat the dictate-send cycle but one individual user (Figure 6.2) prefers a longer chain in which read, dictate and send are followed by another cycle of read-dictate-send.

In other words, the general pattern seems to be to read the messages first and then dictate and send messages in the order they occur in the mailbox, while one particular user prefers to deal with each of the messages one at the time, reading, dictating and sending a reply to the message as she goes through the mailbox. On the other hand, the particular user also tended to be more certain about the message after having dictated it: she either sent it or cancelled the action, whereas the others usually used the listen-option to confirm that the dictated message was as intended.

Another difference that can be visualised is that users seem to have fixed habits. Some users exploit a limited part of the functionality available, although they use the system quite a lot. For instance, two of the users called to the system almost the same number of times during experiment, but the other user used only 13

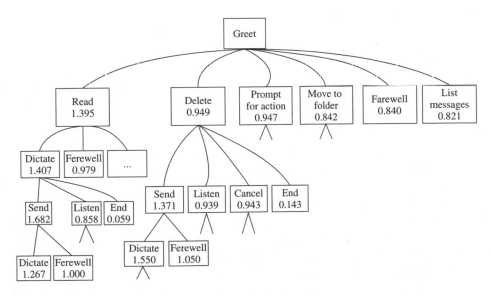

Figure 6.1   Visualisation of the possible dialogue paths for all partners. Jokinen *et al*. (2002)

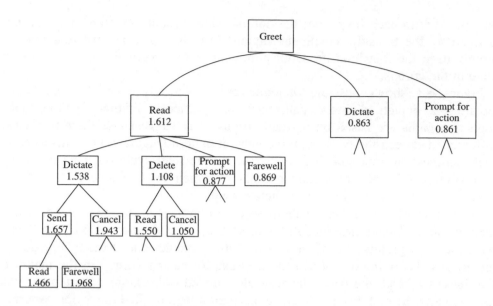

Figure 6.2 Visualisation of the possible dialogue paths for the dialogue partner A. Jokinen *et al.* (2002)

different three-state dialogue chains in his dialogues, whereas the other user had up to 50 different dialogue chains. This kind of variation in the individual users' interaction habits is the knowledge that an intelligent computer agent is to learn and store in the user model, and which can then be exploited in determining the user's individual habits and preferences on-line.

6.2 Cognitive Systems and Group Intelligence

As discussed in Section 3.1.1, the physical and biological levels of communication concern information exchange on sub-symbolic subconscious and automatic levels where chemical and physical causal laws operate. Neural mechanisms form the basis for action, perception, and cognitive processes, and are thus necessary for enabling more conscious communication and interpretation of the observations and sensations of the agent. In this context, the so-called mirror neurons have attracted a lot of interest as they do not only activate during the agent's specific actions (e.g. hand movements) but also respond to the agent's perception of the action, e.g. they activate in the presentation of hand movement, without any overt hand movement by the agent (Gallese *et al.* 1996; Rizzolatti & Craighero, 2004). Mirror neurons suggest that the production and interpretation of actions are related to each other on a neuronal level, and they can be considered as part of a larger cortical system that enables individuals to understand the behaviour of others. By matching the observations of the other agent's action with the execution of one's own actions, the agents develop understanding of their environment and the other agent's behaviour; their learning is based on their experience through their own sensorimotor capacities. From the communication point of view, alignment and cooperation (see Section 3.1.3) can be seen as having their basis in this kind of neural mirroring, as well as such social aspects of cognition as empathy and ethics (Thompson, 2001; Thagard, 2007).

Although the agent's messaging on the neuro-biological level of communication cannot be described as directed behaviour in the same sense as conscious action and communication, the emerging cortical patterns come to exhibit the types of concepts and schemas that form the foundation for human cognition and consciousness. The systematic relationships between observations and actions that apply to human higher-level activities on psychological and social levels of communication (and which are governed by social obligations and general cooperation principles), must thus emerge as a result of the agent's exposition to certain kind of incoming data which fires certain connections among the neurons and cascades up in the network of neurons, providing the basis for grounding language in experience and shared meaning. The relation from sensory information to cognitive categories is not straightforward, but active and large-scale research is going on, concerning cognitive functions and brain regions (Barsalou, 1999; Arbib, 2003), image schemas as cognitive representation for thought (Mandler, 2004), language learning, and evolution (Tomasello, 1992; Steels, 2003; 2004).

Due to space limits we cannot go into details of the cognitive linguistics and language evolution research. However, it is good to mention the work by Elman *et al.* (1996) who study "innateness" of the human language ability, and discuss the mechanisms that underlie complex behaviours in a "biologically oriented conncectionism". They define three important constraints for language learning and cognition: the architecture, the representation and the time. The architecture defines constraints for development along a number of dimensions and granularity levels although it is possible that the architecture itself evolves, too. The representation defines the way in which the patterns and higher-level concepts can be expressed and thus also defines the constraints for development. Time, finally, is the ultimate element that defines the evolution itself: the changes that take place through time accumulate in the architecture and representation, and a certain kind of development is possible only at a certain time when the appropriate architecture and representation are available. The constraints at these levels interact with each other, and the observable behaviours arise as a result of the interaction. Concepts and cognitive categories have their neuronal basis in the patterns that are formed by the brain activations and the form of which is determined by the connections between the neurons. However, the new sensory data input modifies neuronal connections: the brain is in a constant process of change and new things are learnt. Learning requires a lot of new data input, however, so that new connections can replace the established ones. On the other hand, the old connections direct us to see the world in a particular way: we expect certain patterns that reinforce our previous experience, and we interpret the data input so that it fits to the existing patterns. The discrepancy between new data and an established pattern should be quite large for new learning to take place, and new established patterns to become formed on the basis of the inconsistent data flows.

If this view of learning and change on the neurobiological level is expanded towards learning cooperative communication and social interaction, we can assume that the three important aspects are involved again: architecture, representation and time. It may be difficult to imagine an analogy for these aspects in social communication, since ultimately they refer to the individual brains that process language. However, taking into account the fact that language is a social as well as individual phenomenon, we can try to find out if the framework established for an individual to interact with the environment, would also analogously work for a group of individuals trying to coordinate their actions and using language to communicate with each other.

Given that the individuals in a group play a similar role of "processing units" as neurons in a neural network, an analogy for the architecture can be found in the social groups of cooperating agents. The groups have their own sociodynamics which constrain the individual agents' behaviour. For instance, the group imposes social norms and principles on its members (which, as we argue in this book, can be defined through the agents' rational coordinated action, see Section 3). Although the norms are not causal rules that determine the agent's course of action, they are obligating requirements that can constrain the agent's behaviour significantly. The norms depend on the agents' roles and activities within the group, i.e. they are determined by the agent's position in the social hierarchy within a certain activity, and can be compared with the position and connections of a neuron in a connectionist network.

In the distributed AI the group dynamics have been modelled with respect to the coherence and coordination of multiagent systems. Besides the game-theoretic approach (see e.g. Gmytrasiewicz and Durfee, 2000; Rasmusen, 1994), the biologically inspired computing paradigms such as Ant-Colony Optimization or Swarm Intelligence (see an overview in Bonabeau, Dorigo & Theraulaz, 1999) offer techniques to model such dynamics. The latter emphasise the emergent nature of problem solving behaviour as encountered in the colonies of ants and swarms of bees, resulting from the collective activity of the simple creatures situated in a dynamically changing environment and exhibiting no central control. Social behaviour of large populations can thus be modelled as an optimization problem where agents interact locally with their environment but their seemingly random behaviour causes global patterns to emerge. The artificial ants are not intelligent agents, however, in the same sense as we have studied agents in this book; rather, they are simple elements that exhibit intelligent behaviour collectively. From the point of view of social dynamics, the individual intelligence does not play a main role here: the complex behaviour of a swarm system is based on collective reaction and adaptation to the dynamic environment rather than on the individual's planning and action coordination. The swarm intelligence thus operates beyond the individual's direct control although it may have its origin in the agent's actions. Considering intelligent and rational human agents, we can say that swarm intelligence falls within the agent's social intelligence, while deliberation on how it affects coordination of tasks and actions is part of the agent's social skills and interaction ability. In other words, the agent's participation in the "swarm" is often conditional on a more or less conscious decision to commit oneself to the shared goals, rather than a built-in reaction to the environment as it is with ants and bees (see discussion on teamwork and cooperation in Section 2.2.5 and Section 3.1.3)

Related to this, another important difference between multi-agent systems and the agent groups and communities that we consider in this book is the emphasis on communication. While in multi-agent systems the goal is usually to operate with a minimum communication overhead, the groups of intelligent agents need to communicate in order to coordinate their intentions. The autonomous agents may have contradictory individual intentions with each other, and their goals do not necessarily accord with the goals of the group they belong to, so they need to plan and negotiate with the other agents in a deliberate manner, about how to achieve their goals. In these situations, communication is necessary for learning to know the constraints and preferences of other agents: in other words, communication is not an overhead but a necessity. The most natural means of communication is natural language which provides a structure and meaning repertoire rich enough for such abstract deliberations. In fact, it is possible to claim that the reward for human multi-agent groups in developing a means of communication such as the natural language is especially to enable rich intentional communication between the individuals, so as to better coordinate and resolve conflicts concerning individual and shared tasks (for hunting and food gathering). Moreover, often the agents enjoy communication without any other overt purpose than keeping contact with each other, and in these situations the language is a means of exchanging information on topics that do not presuppose efficient and task-oriented actions.

Concerning representations, we can extend the analogy from neuronal activation patterns to cultural representations and collective norms. These exhibit prevailing social patterns and normative constraints on the individual's and group behaviours, and thus function over multi-agent systems (groups, communities, nations) in a similar way as the fine-grained micro-level patterns of cortical activations in the brain. Such macro-level knowledge as exhibited in the human cultural inheritance, history, art, folklore, collective memory, etc., represent the understanding of the activities produced by groups and communities, and can thus be compared with cognitive categories produced as a result of neuronal activity.

Finally, as in the case of evolution of individuals, time is an important factor in the development, specialization, and maturation of the agent groups, communities, and cultures. However, although it seems obvious that the timing of input plays a crucial role also on this level, the absence of felicitous stimuli may not have as vital consequences in the development of complex social behaviours as it has in the ontogeny of physiological organs.

6.3   Interaction and Affordance

Intertwined with the issue of desirability of adaptation (as a property of computer agents) is the question of how adaptation should be modelled and how it should take place in human-computer interactions. Is the user to be in charge of adaptation or are there cases where automatic adaptation would be desirable? In the ideal case, of course, automatic adaptation would support intuitive interfaces which adapt to the various users providing a natural interaction without the user even noticing that something special has taken place in the system's behaviour. After all, this is what happens constantly in human-human communication: adaptation to the partner is so automatic that we don't necessarily notice that it has happened; rather it leaves only a positive feeling towards a nice conversational partner with whom it was a pleasure to converse. The opposing view sees here a seed for mistrust. In a similar way to that in which humans can be unreliable and lie, the adaptive dialogue system can also lie. How are we going to be sure that the information given to the user is truthful and undistorted? Perhaps it is the complex interaction between the various levels of interpretation and reasoning that causes fear rather than lack of truthfulness: interaction with something which one does not know and cannot control is a threat as such. However, as with human-human communication, also in interactions with computer agents, the users base their cooperation on mutual trust. Interaction presupposes that the partners behave according to the principles of Ideal Cooperation: the partners are rational agents who normally act reliably and provide truthful information. Another question is then how this kind of trust is created and maintained within the communication. However, once the trust is gone, it is very difficult to remedy it.

One of the main questions in the design and development of natural and cooperative interactive computer agents, is what it actually means to communicate in a natural way in different situations. In the context of intelligent computer agents the adjective "natural" should not only refer to the agent's ability to use natural language, verbal and non-verbal communication, but to support functionality that the user finds intuitive and easy. For instance, gesturing is natural when pointing and giving spatial information, while face and eye-gazing express emotions and focus attention best. Speech, on the other hand, is not suitable if one is concerned about disturbing others or if there are privacy issues, or if the enablement for speech communication is not fulfilled at all. Consequently, we can say that the interactive agent should *afford* natural interaction.

The concept of affordance was originally introduced by Gibson (1979) who used it in his visual perception studies to denote the properties of the world that the actors can act upon. It was brought to product design and human-computer interaction by Norman (1988) to capture the essence of good design, i.e. a well-designed artefact in itself suggests its most natural use. Affordance can be divided into real and perceived affordances depending on whether we talk about the uncountable number of actionable relationships between the actor and the world, or about the action possibilities that are perceived by the user, respectively. In HCI, affordance is usually related to the perceived affordances, referring to the properties of the object that readily suggest to the user the appropriate ways to use the artifact. The smart technology environment should thus afford various natural interaction techniques, i.e interfaces should lend themselves to natural use without the users needing to think and reason how the interaction should take place in order to get the task completed.

In complex interactive environments, where the tasks are abstract and require coordination of action for a shared understanding, the most natural means of interaction is language communication, including both verbal and non-verbal modalities. Analogously, when interacting with an intelligent computer agent, the dialogue interface in itself suggests natural language as the most natural means of interaction; in other words, the dialogue interface thus readily affords natural communication.

7

Conclusions and Future Views

We tend to think of inanimate objects as tools that can be used to perform particular tasks. Computers, originally developed as calculating machines to sort out vast amounts of numerical information can be regarded as such tools: their functioning is basically designed so as to adequately to fit with the requirements of the task at hand. However, the development of technology has also changed our views of the computer: it is not only a tool but a complex system whose manipulation is not necessarily a step wise procedure of commands, but resembles natural language communication.

In this book I have argued that our view of computers has changed due to their new interaction and natural language capabilities. Computer systems are regarded as interactive agents, and the standards of interaction approach more natural and intuitive style: the human factor in a robust and adequate interface is the ability to use symbolic natural language.

The Constructive Dialogue Model approach regards communicating partners as rational agents. The crucial aspects of natural, intuitive communication deal with the goals and intentions that the agents want to communicate to each other and, as a result, can also share between them (with the effect that the agents mutually believe that they share these goals). The rationality of the agents thus manifests itself in the manner in which the agents act and reason about the communicative situation, and present the new information to the partner. It must be pointed out, however, that rationality is a relative concept measurable with respect to a particular interaction. It is a property assigned by the agent to their dialogue partners on the basis of observing and evaluating the partners' actions, and it is reinforced through the positive feedback of the successful achievement of similar goals in other situations.

In searching for the definition of rational action, our view-point has been that of communication in general: agents are engaged in interactive communicative situations and their actions are governed by social obligations. Communication is fundamentally cooperative action whereby the agents respond to the changes in the context in which the interaction takes place. Since the agents evaluate the changes with respect to their own goals and motivations, and are also obliged to

Constructive Dialogue Modelling   Kristiina Jokinen
© 2009 John Wiley & Sons, Ltd

report on the result of this evaluation back to the partner, rationality emerges from this mutual activity in the form of the agent's compliance with the requirements of general communicative principles. Rationality is not a static property which the agents may or may not possess, but an effect related to the gradual convergence of the goals set for the interaction. The observation that rationality is in fact the perceived rationality of an agent's actions, leads us also to a straightforward definition of irrational behaviour: if the agent's interaction with the partner does not appear to converge upon a certain shared conclusion, the agent's behaviour is described as irrational, lacking the basic characteristics of a motivated and competent agent.

Following Allwood's notion of Ideal Cooperation, which states that communication is a cooperative activity whereby they share a goal, show cognitive and ethical consideration towards the partner and the situation, and believe that the partner follows the same principles in their reasoning. The ethical dimension, which obliges the agent to act so as not to prevent other agents from achieving their goals, distinguishes Ideal Cooperation from most theories of rational agency, and provides an important counter-force to the epistemic rationality which deals with the agent's knowledge of what is a rational way to plan and maximise one's returns. An interesting topic for future research in the ubicom context is how the agents learn to rank (communicative) actions as appropriate and ethically acceptable, and how the agents' view-point, background assumptions, goals and social obligations affect this process.

It is also important to study what kind of impact such intelligent systems have for users and social environment as a whole. Besides increasing our knowledge of the enablement and constraints for effective natural communication, this will also contribute to our understanding of the impact of the new technology on society and people's lives in general.

First of all, social interaction among people changes. The internet seems to develop from a technical structure to a platform which converges all kind of knowledge into digital form and also becomes a global meeting point, where users can download and receive data from the other internet users, and share own (digital) data with friends and colleagues and other members of the community. Virtual communities are thus rapidly formed of users with similar interests: interaction is rapid but not necessarily face to face, and true identities of the participants may be hidden behind different roles. Email, blogs, web navigation and browsing provide user-controlled ways to exchange information, and internet phones, videoconferencing and chat rooms enable real time interaction. In novel types of virtual worlds users can even create their identities and live in a world of other avatars (Habbottotel, SecondLife, OnlineLife). Mobile communication possibilities (GPRS, GPS, wireless networks) allow communication anywhere anytime.

This kind of on-line communication presupposes the off-line processing of vast amounts of data which consist of various types of digital data such as texts, music, photos, videos and emails. The organisation of data should be automatic and fast, and allow for human intervention in directing and guiding the process according to individual preferences and needs, as well as in retrieving information according to some topical principles which relates to the conversion and the topic that the user is talking about. The situation is a typical interface agent task where the computer agent assists the user with the interaction with the application. In this particular case, monitoring of the conversation and recognition of the dynamically changing topic are of vital importance, in order for the system to provide quick responses and appropriate information.

Secondly, the interaction with the system changes. As argued in this book, users tend to regard the system in more human-like terms than the computer-as-a-tool metaphor would warrant. With the development of the ubicom paradigm, an increasing number of interactive applications will also appear in our environment, all competing with each other from the user's attention. However, interactions with context-aware systems are not only simple notifications of apparently interesting and important information, but they also function as summoning the user's attention. From the user's point of view, the situation may resemble a telephone call which requires answering and possibly a longer interaction (cf. studies on interaction patterns and starting of conversations, Sacks 1992). As the user's current task is interrupted and the user is summoned to respond, the user's expectations for natural and intuitive communication may also be summoned. On the other hand, in order to get the message through, the application needs to be aware of the user's current situation, take into account the user's familiarity with the application and her trust in the information source, as well as consider possible consequences of the incoming information on the user's current situation. It is thus important that intelligent agents are equipped with conversational capabilities, i.e. interaction management should take into account human communicative principles concerning cooperation, obligations, rights, and trust.

There appears to be a fundamental asymmetry between the user and the system, however. The question has especially concerned bi-directionality, or the difference between the system's language interpretation and generation capabilities. According to a traditional view, the practical service system should at least understand all of the utterances that it is capable of producing, but for smooth communication, it should understand more structures and words than it can produce, since the users should not be restricted by the system's limited language capability. Communication abilities are, however, bidirectional. In accordance with the Ideal Cooperation, the agent interprets and reacts to the partner's contribution, and the reaction functions as the starting point for the partner's interpretation. The agent assumes that the partner follows the same logic and principles as the agent and their interaction thus resembles a game where the moves are performed according to the rules of the game in order to achieve the goals.

Finally, it should be noted that flexible dialogue management also requires a better understanding of human perception and cognitive processing. Even if we leave the philosophical problems of consciousness and thinking machines aside, the models for symbolic learning systems benefit from the investigations into neural and brain-like computation: general architectures, learning algorithms and memory organisation form the basis for domain-dependent representations and attentional state, from which conceptualisation and language learning can emerge as a result of interactions of constraints on various levels (Elman et al., 1996; Ellman, 1999).

It may, of course, be that the future computational systems exceed human processing capability by implementing an intelligent algorithm that has no resemblance to the human way of processing at all. In this case, the systems exhibit a new type of "alien" intelligence which may not only seem unfamiliar but also threatening to the users. In principle, chess playing programs and intelligent robot cars already belong to this class, as they autonomously perform tasks that have previously been performed by humans only. It is also possible to imagine a future society where silicon avatars live together with humans, taking part in the activities as equal agents as well as friends and communicating entities (Kurzweil, 1999). However, we do not easily allow ourselves to consider a non-living machine as an equal communicating partner, nor do we think that a machine can have similar properties, emotions and intentions as humans do, e.g. the ability to show compassion, or to lie (cf. also the "Uncanny Valley" hypothesis of "almost human" android robots, discussed in Chapter 1). It has also been pointed out that the concept of equality presupposes autonomy, which is a problematic concept in itself. From the point of view of rational agency, autonomy concerns voluntary acting and independence, and we are faced with the philosophical questions of freedom of will. On the other hand, the actions of the computer agent can also be seen as simulations of what we as human agents interpret as intelligent behaviour (cf. Reeves & Nass, 1996). In other words, once we enter into an interaction with a computer agent, the human natural reaction patterns are activated automatically, and we perceive the agent's reactions as similar reactions to those that an intelligent human partner would have exhibited with a certain psychological and cognitive motivation. Without going too much into philosophical problems concerning thinking machines, consciousness and free will, we can ask if similar cognitive activities that are typical in natural interaction between humans, can emerge in computer agents and in agent populations consisting of computers. As the book has put forward an approach which takes into account rich possibilities for interaction, the answer is "yes". However, it has not been the aim of this book to argue for the simulation of human intelligence as such, with aspects that cannot be considered optimal behaviour. To the contrary, the intention has been to support a view that, by furnishing the computer with capabilities that seem to be equivalent to human intellectual capabilities (e.g. chess playing, intelligent car driving), even surpassing them in some aspects (fast calculations and search), we also need to have a closer look at the system's communicative competence, ie its interaction capability with its environment, especially with human users. An intelligent system should also possess capabilities that enable it to interact with humans in a natural and intuitive way. This means that computer agents should also at least be able to exchange information about their internal state and knowledge so as to be able to construct a shared understanding of how to coordinate tasks and actions in order to achieve individual and mutual goals. The kind of "minimum rationality" does not presuppose that we reinvent the human, but it supports the more modest goal of building intelligent systems that possess enough communicative intelligence so as to assist human users in making informed decisions about their complex environment. If the computer agent can exhibit capability to follow human logic and communication style, it is likely to be perceived as a user-friendly interface to the digital world, capable of attending the user's needs and acting in a rational cooperative manner. The computer agent's ability to conform to the general principles of rational cooperative communication will thus contribute to its positive evaluation as a useful practical service system, and will also support its appearance as being in possession of such cognitive capabilities that are typical for intelligent agents.

The main challenge for 21st century communication technology seems to be natural, i.e. intuitive and natural language based interaction between humans and computer agents. The following appear to be the main issues in integrating language and interaction research with advanced technology, to support the "minimal communicative intelligence":

- *Knowledge.* Type and representation of knowledge, and the nature and number of reasoning rules are important prerequisities for intelligent communicative behaviour. Data-driven statistical and machine learning techniques are used for concept and rule-extraction but need to be supported by symbolic reasoning, and incorporated into interactive applications.
- *Adaptation.* An intelligent communicating system should be capable of learning and adapting itself to new tasks, situations, users, language expressions, etc. Adaptive methods are also required in order to enable systems to evolve over time so as to meet dynamically changing needs of the users.
- *Responsiveness.* The intelligence of a computer agent is shown in its capability to handle processing errors and incomplete information. Besides context modelling and exploiting of expectations, this requires that processing units and their interaction are specified in a way that allows for ambiguous information exchange between the modules.
- *Multimodal and nonverbal interaction.* Sophisticated models of interaction take multimodality into account and verbal and non-verbal communication into account, and allow effective use of gestures, body posture, and facial expression together with the interpretation and generation of verbal utterances. The system's communicative repertoire is extended with multimodal technology which also caters for new interface possibilities for various user groups with special needs.

- *Neuro-cognitive processing models*. Advances in the research on human cognitive processing has brought important information about the subsymbolic and sensorimotoric basis of communication (e.g. mirror neurons). Integration of such models into intelligent computer agents is a challenging task.

In general, the "minimal communicative intelligence" requires a theory and formalisation of human communication related to the constraints on representations, architectures and processing through which the repetitive communicative patterns in processing units emerge.

References

Allen, J., Byron, D., Dzikovska, M., Ferguson, G., Galescu, L. and Stent, A. 2000. An Architecture for a Generic Dialog Shell. *Natural Language Engineering* **6** (3): 1–16.

Allen, J. F. and Perrault, C. R. 1980. Analyzing Intention in Utterances. *Artificial Intelligence* **15** (3): 143–78.

Allen, J. F., Miller, B. W., Ringger, E. K. and Sikorski, T. 1996. A Robust System for Natural Spoken Dialogue. *Proceedings of the 1996 Annual Meeting of the Association for Computational Linguistics (ACL'96)*, 62–70.

Allen, J. F., Schubert, L. K., Ferguson, G., Heeman, P., Hwang, C. H., Kato, T., Light, M., Martin, N. G., Miller, B. W. Poesio, M., Traum, D. R. 1995. The TRAINS Project: A Case Study in Building a Conversational Planning Agent. *Journal of Experimental and Theoretical AI*, **7**: 7–48. Also available as TRAINS Technical Note 94-3 and Technical Report 532, Computer Science Dept., University of Rochester, September 1994.

Allwood, J. 1976. *Linguistic Communication as Action and Cooperation*, Gothenburg Monographs in Linguistics 2, University of Göteborg, Department of Linguistics.

Allwood, J. 1977. A Critical Look at Speech Act Theory. In Dahl, Ö. (Ed.) *Logic, Pragmatics, and Grammar*. Studentlitteratur. Lund, pp. 53–69.

Allwood, J. 1984. On Relevance in Spoken Interaction. In Kjellmer, G. and Bäckman, S. (Eds.) *Papers on Language and Literature*, Acta Universitatis Gothoburgensis. pp. 18–35.

Allwood, J. 1992. On Dialogue Cohesion. Gothenburg Papers in Theoretical Linguistics 65, University of Göteborg, Dept of Linguistics. Also in Heltoft, L. and Haberland, H. (Eds.), (1996) *Papers from the Thirteenth Scandinavian Conference of Linguistics*, Roskilde University, Dept of Languages and Culture.

Allwood, J. 1993. Feedback in Second Language Acquisition. In Perdue, C. (Ed.) *Adult Language Acquisition. Cross Linguistic Perspectives, Vol. II*. Cambridge: Cambridge University Press, pp. 196–235.

Allwood, J. 1994. Obligations and Options in Dialogue. *Think Quarterly* **3**: 9–18.

Allwood, J. 1995. *An Activity Based Approach to Pragmatics*. Gothenburg Papers in Theoretical Linguistics 76, Göteborg University, Department of Linguistics.

Allwood, J. 2000. An Activity Based Approach to Pragmatics. In Bunt, H., Black, W. J. (Eds.) *Abduction, Belief and Context in Dialogue: Studies in Computational Pragmatics*. Amsterdam: John Benjamins, pp. 47–80.

Allwood, J. 2002. Bodily Communication – Dimensions of Expression and Content. In Granström, B., House, D., and Karlsson, I. (Eds.) *Multimodality in Language and Speech Systems*. Kluwer. pp. 7–26.

Allwood, J. 2007. *Activity Based Studies in Linguistic Interaction*. Gothenburg Papers in Theoretical Linguistics 93. Göteborg University, Department of Linguistics.

Allwood, J., Cerrato, L., Jokinen, K., Navarretta, C. and Paggio, P. 2005. The MUMIN Annotation Scheme for Feedback, Turn Management and Sequencing. *Proceedings of the 2nd Nordic Conference on Multimodal Communication*, Göteborg, Sweden. http://cst.dk/mumin/ workshop200406/MUMIN-coding-scheme-v1.3.doc

Allwood, J, Cerrato, L., Jokinen, K., Navarretta, K., and Paggio, P. 2007. The MUMIN Coding Scheme for the Annotation of Feedback, Turn Management and Sequencing Phenomena. In Martin, J. C., Paggio, P., Kuenlein, P., Stiefelhagen, R., and Pianesi, F. (Eds.) *Multimodal corpora for modelling human multimodal behaviour*. Special issue of the *International Journal of Language Resources and Evaluation*, **41** (3–4), 273–87. http://www.springer.com/journal/10579/.

Allwood, J., Nivre, J., Ahlsén, E. 1992. On the Semantics and Pragmatics of Linguistic Feedback. *Journal of Semantics*, **9**: 1–29.

Allwood, J., Traum, D., Jokinen, K. 2000 Cooperation, Dialogue and Ethics. *International Journal of Human-Computer Studies*, **53**: 871–914. doi 10.1006/ijhc.2000.0425. Available online at http://www.idealibrary.com.on.

Anderson, A. H., Bader, M., Bard, E. G., Boyle, E., Doherty, G., Garrod, S., Isard, S., Kowtko, J., McAllister, J., Miller, J., Sotillo, C., Thompson, H. S. and Weinert, R. 1991. The HCRC Map Task Corpus. *Language and Speech* **34**(4): 351–366.

André, E. and Pelachaud, C. (forthcoming) Interacting with Embodied Conversational Agents. In Jokinen, K. and Cheng, F. (Eds.) *New Trends in Speech-based Interactive Systems*. Springer Publishers.

Appelt, D. E. 1985. *Planning English Sentences*. Cambridge: Cambridge University Press.

Arbib, M. 2003. The Evolving Mirror System: A Neural Basis for Language Readiness. In Christiansen, M. and Kirby, S. (Eds.) *Language Evolution*. pp. 182–200. Oxford: Oxford University Press.

Aust, H., Oerder, M., Seide, F. and Steinbiss, V. 1995. The Philips Automatic Train Timetable Information System. *Speech Communication* **17**: 249–262.

Austin, J. L. 1962. *How to do Things with Words*. Oxford; Clarendon Press.

Axelrod, R. 1984. *Evolution of Cooperation*. New York: Basic Books. Revised edition Perseus Books Group, 2006.

Baddeley, A. D. 1992. Working Memory. *Science*, **255**: 556–9.

Ballim, A., Wilks, Y. 1991. *Artificial Believers*. Hillsdale, New Jersey: Lawrence Erlbaum Associates.

Barsalou, L. W. 1999 Perceptual Symbol Systems. *Behavioral and Brain Sciences*, **22**: 577–660.

Barto, A. and Sutton, R. 1993. *Reinforcement Learning: An Introduction*. Cambridge, Mass.: MIT Press.

Bateman, J. A. 1995. On the Relationship between Ontology Construction and Natural Language: a SocioSemiotic View. *International Journal of Human-Computer Studies*, **43**: 929–44.

Becker, C. and Wachsmuth, I. 2006. Playing the Cards Game SkipBo against an Emotional Max. In Reichardt, D., Levi, P. and Meyer, J.-J. C. (Eds.), *Proceedings of the 1st Workshop on Emotion and Computing*, (p. 65), Bremen.

Behnke, S., Stückler, J., Strasdat, H., Schreiber, M. 2007. Hierarchical Reactive Control for Soccer Playing Humanoid Robots. In M. Hackel (Ed.), *Humanoid Robots, Human-like Machines*, I-Tech. pp. 625–642.

Bennewitz, M., Faber, F., Joho, D., Behnke, S. 2007. Fritz – A Humanoid Communication Robot. Proceedings of the 16th IEEE International Symposium on Robot and Human Interactive Communication (RO-MAN).

Berners-Lee, T. (with Mark Fischetti) 1997. *Weaving the Web. The Original Design and Ultimate Destiny of the World Wide Web*. San Francisco: Harper.

Bernsen, N. O., H. Dybkjaer and L. Dybkjaer 1998. *Designing Interactive Speech Systems. From First Ideas to User Testing*. London: Springer Verlag.

Bernsen, N. O. and Dybkjaer L. 2002. DISC dialogue engineering best practice guide to dialogue management. NISLab, University of Southern Denmark. http://www.disc2.dk/.

Berthold, A., Jameson, A. 1999. Interpreting Symptoms of Cognitive Load in Speech Input. In Kay, J. ed. *Proceedings of the 7th International Conference on User Modelling (UM99)*, Wien: Springer, pp. 235–244.

Beun, R. J. and Cremers, A. H. M. 1998 Object Reference in a Shared Domain of Conversation. *Pragmatics and Cognition* **6**(1/2): 121–52.

Bilange, E. 1991. A Task Independent Oral Dialogue Model. *Proceedings of the 5th Meeting of the European Chapter of the Association for Computational Linguistics*, pp. 83–8.

Bilange, E., Magadur, J.-Y. 1998 A Robust Approach For Handling Oral Dialogues. *Proceedings of COLING 1992*, pp. 799–805.

Billard, A. and Dautenhahn, K. 1999. Experiments in Learning by Imitation – Grounding and Use of Communication in Robotic Agents. *Adaptive Behavior*, **7**(3/4): 415–38.

Bird, S., Liberman, M. 1999 A Formal Framework for Linguistic Annotation, *Speech Communication*, **33**: 23–60.

Black, W. J., Bunt, H. C., Dols, F. J. H., Donzella, C., Ferrari, G., Haidan, R., Imlah, W. G., Jokinen, K., Lager, T., Lancel, J-M., Nivre, J., Sabah, G. and Wachtel, T. 1991. *A Pragmatics-Based Language Understanding System*. PLUS Deliverable D1.2.

Blaylock, N., Allen, J., Ferguson, G. 2002. Synchronization in an Asynchronous Agent-based Architecture for Dialogue Systems. In Jokinen, K. and S. McRoy (Eds.) *Proceedings of the 3rd SIGDial workshop on Discourse and Dialogue*. Philadelphia, U.S., pp. 1–10.

Bohus, D., Rudnicky, A. I. 2003. RavenClaw: Dialog Management using Hierarchical Task Decomposition and an Expectation Agenda, *Eurospeech 2003*, pp. 597–600.

Bolt, R. A. 1980. Put-that-there: Voice and gesture at the graphic interface. *Computer Graphics*, **14**(3): 262–70.

Bonabeau, E., Dorigo, M., Theraulaz, G. 1999. *Swarm Intelligence: From Natural to Artificial Systems*. Oxford: Oxford University Press.

Bos, J. Klein, E. Oka, T. 2003. Meaningful Conversation with a Mobile Robot. *Proceedings of the 10th Conference of the European Chapter of the Association for Computational Linguistics (EACL 2003)*, pp. 71–74.

Bos, J., Klein, E., Lemon, O. Oka, T. (2003). DIPPER: Description and Formalisation of an Information-State Update Dialogue Sysem Architecture. In 4th SIGdial Workshop on Discourse and Dialogue, pp. 115–24, Sapporo.

Bratman, M., Israel, D., Pollack, M. E. 1988. Plans and Resource-Bounded Practical Reasoning. *Computational Intelligence*, **4**(4): 349–55.

Breazeal, C. (2002). *Designing Sociable Robots*, MIT Press, Cambridge, MA.

Brooks, R. A. 1986. A Robust Layered Control System for a Mobile Robot. *IEEE Journal of Robotics and Automation*, **2**(1): 14–23.

Brooks, R. A. 1991. Intelligence Without Reason, *Proceedings of 12th International Joint Conference on Artificial Intelligence*, Sydney, Australia, pp. 569–595.

Brown, P. Levinson, S. 1987. *Politeness: Some Universals in Language Usage*. Cambridge: Cambridge University Press.

Brown, P., Levinson, S. C. 1999 [1987]. Politeness: Some Universals in Language Usage. In Jaworski, A., and Coupland, N. (Eds.) *The Discourse Reader*, pp. 321–335. London: Routledge.

Brown-Schmidt, S. Tanenhaus, M. 2004. Priming and Alignment: Mechanism or Consequence? *Behavioral and Brain Sciences* **27**: 193–4.

Brusilovsky, P., Maybury, M. T. 2002. From Adaptive Hypermedia to the Adaptive Web. *Communications of the ACM* **45**(5): 30–3.

Buchanan, B., Shortliffe, E. 1984 *Rule Based Expert Systems*. Reading, Mass.: Addison Wesley.

Bunt, H. C. 1990. DIT – Dynamic Interpretation in Text and Dialogue. In Kálmán, L. and Pólos, L. (Eds.) *Papers from the Second Symposium on Language and Logic*. Budapest: Akademiai Kiadó.

Bunt, H. C. 2000. Dynamic Interpretation and Dialogue Theory. In Taylor, M. M., Néel, F. and Bouwhuis, D. G., (Eds.) *The Structure of Multimodal Dialogue II*, pp. 139–166. Amsterdam: John Benjamins.

Bunt, H. C. 2005. A Framework for Dialogue Act Specification. *Proceedings of the Fourth Workshop on Multimodal Semantic Representation (ACL-SIGSEM and ISO TC37/SC4)*, Tilburg.

Bunt, H., Girard, Y. 2005. Designing an open, multidimensional dialogue act taxonomy. Proceedings of DIALOR'05, 37–44.

Burge, C., Doran, C., Gernter, A., Gregorowicz, A., Harper, L., Korb, J., Loehr, D. 2003. Dialogue complexity with portability? Research Directions for the Information State Approach. *Proceedings of the HLT-NAACL/NSF Human Language Technology Conference workshop on Research Directions in Dialogue Processing*, Edmonton, Canada. Software available at: http://midiki.sourceforge.net/.

Campbell, N. 2007. On the Use of Nonverbal Speech Sounds in Human Communication. In N. Campbell (Ed.), *Verbal and Nonverbal Communication Behaviors*. Lecture Notes in Articicial Intelligence 4775, pp. 117–128). New York: Springer.

Carberry, S. 1990. *Plan Recognition in Natural Language Dialogue*. Cambridge, Mass.: MIT Press.

Carletta, J. 1992. Planning to Fail, not Failing to Plan: Risk-taking and Recovery in Task-oriented Dialogue. *Proceedings of COLING*, pp. 896–900, Nantes.

Carletta, J. 1996. Assessing Agreement on Classification Tasks: The Kappa Statistics. *Computational Linguistics*, **22**(2): 249–254.

Carletta, J., Dahlbäck, N., Reithinger, N., Walker, M. (Eds.) 1997. Standards for Dialogue Coding in Natural Language Processing. Dagstuhl-Seminar Report 167.

Carletta, J., Isard, A., Isard, S., Kowtko, J., Doherty-Sneddon, G., Anderson, A. 1996. HCRC Dialogue Structure Coding Manual. Tech.Rep. HCRC TR-82, Human Communication Research Centre, University of Edinburgh, Edinburgh, Scotland.

Carlson R. 1996. The Dialog Component in the Waxholm System. In LuperFoy, S., Nijholt, A., and Veldhuijzen van Zanten, G. (Eds.) *Proceedings of Twente Workshop on Language Technology. Dialogue Management in Natural Language Systems (TWLT 11)*, pp. 209–18.

Cassell, J., Nakano, Y. I., Bickmore, T. W., Sidner, C. L., Rich, C. 2001a. Non-verbal Cues for Discourse Structure. *Proceedings of ACL*, pp. 106–115.

Cassell, J. Vilhalmsson, H., Bickmore, T. 2001b. BEAT: the Behaviour Expression Animation Toolkit, *Proceedings of SIGGRAPH '01*, pp. 477–86, Los Angeles, CA.

Cassell, J., Sullivan, J., Prevost, S., Churchill, E. (Eds.) 2003. *Embodied Conversational Agents*. Cambridge, Mass.: MIT Press.

Cawsey, A. 1993. *Explanation and Interaction: the Computer Generation of Explanatory Dialogues*. Cambridge, Mass.: MIT Press.

Cawsey, A., Galliers, J., Logan. B., Reece, S., and Spärck-Jones, K. 1993. Revising Beliefs and Intentions: A Unified Framework for Agent Interaction. *Proceedings of the 9th Biennial Conference of the Society for the Study of Artificial Intelligence and Simulation of Behaviour*, pp. 130–139.

Cerrato, L. 2007. Investigating Communicative Feedback Phenomena across Languages and Modalities. PhD Thesis. KTH Computer Science and Communication Department of Speech, Music and Hearing, Stockholm.

Chafe, W. 1976. Givenness, Contrastiveness, Definiteness, Subjects, and Topics. In Li, C. N. (Ed.) *Subject and Topic*, pp. 25–55, New York: Academic Press.

Chafe, W. 1994. *Discourse, Consciousness, and Time*. Chicago/London: University of Chicago Press.

Charniak, E., Goldman, R. P. 1993. A Bayesian Model of Plan Recognition *Artificial Intelligence*, **64**(1): 53–79.

Cherniak, C. 1986. *Minimal Rationality*. Cambridge, Mass.: MIT Press.

Chin, D. 1989. KNOME: Modeling what the User Knows in UC. In Kobsa, A., Wahlster, W. (Eds.) *User Modeling in Dialogue Systems*. Springer-Verlag Berlin, Heidelberg, pp. 74–107.

Chu-Carroll, J. 2000. MIMIC: An Adaptive Mixed Initiative Spoken Dialogue System for Information Queries. *Proceedings of the 6th Conference on Applied Natural Language Processing (ANLP)*, pp. 97–104.

Chu-Carroll, J., Brown, M. K. 1998. An Evidential Model for Tracking Initiative in Collaborative Dialogue Interactions. *User Modeling and User-Adapted Interaction*, **8**(3–4): 215–53.

Chu-Carroll, J., Carberry, S. 1998. Collaborative Response Generation in Planning Dialogues. *Computational Linguistics*, **24**(3): 355–400.

Chu-Carroll, J., Carpenter, B. 1999. Vector-based Natural Language Call Routing. *Computational Linguistics* **25**(3): 256–262.

Chu-Carroll, J., Nickerson, J. 2000. Evaluating Automatic Dialogue Strategy Adaptation for a Spoken Dialogue System. *Proceedings of the First Conference on North American Chapter of the Association for Computational Linguistics (NAACL-00)*, pp. 202–209.

Clark, H. H., Schaefer, E. F. 1989. Contributing to Discourse. *Cognitive Science*, **13**: 259–94.

Clark, H. H., Wilkes-Gibbs, D. 1986. Referring as a Collaborative Process. *Cognition*, **22**: 1–39.

Clark, H. H., Haviland, S. E. 1977. Comprehension and the Given-new Contract. In Freedle, R. O (Ed.), *Discourse Production and Comprehension*, pp. 1–40. Hillsdale, NJ: Erlbaum.

Cohen, P. R., Levesque, H. J. 1990a. Persistence, Intention, and Commitment. In Cohen, P. R., Morgan, J., and Pollack, M. E. (Eds.) *Intentions in Communication*, pp. 33–69. Cambridge, Mass.: MIT Press.

Cohen, P. R., Levesque, H. J. 1990b. Rational Interaction as the Basis for Communication. In Cohen, P. R., Morgan, J. and Pollack, M. E. (Eds.) *Intentions in Communication*, pp. 221–55. Cambridge, Mass.: MIT Press.

Cohen, P. R., Levesque, H. J. 1991. Teamwork. *Nous*, **25**(4): 487–512.

Cohen, P. R., Morgan, J., Pollack, M. (Eds.) 1990. *Intentions in Communication*. Cambridge, Mass: MIT Press.

Cohen, P. R., Perrault, C. R. 1979. Elements of Plan-based Theory of Speech acts. *Cognitive Science*, **3**: 177–212.

Colby, K. M. 1971. Artificial Paranoia. *Artificial Intelligence*, Volume 2.

Cole, R. A., Mariani, J., Uszkoreit, H., Zaenen, A., Zue, V. (Eds.) 1996. Survey of the State of the Art in Human Language Technology. Also available at http://www.cse.ogi.edu/CSLU/HLTSurvey/.

Cooper, R., Larsson, S. 2003. Accommodation and Reaccommodation in Dialogue. In Bäuerle, R., Reyle, U., Zimmerman, E. (Eds.) *Presuppositions and Discourse*. Amsterdam: Elsevier.

Core, M. G., Allen, J. F. 1997. Coding Dialogs with the DAMSL Annotation Scheme. Working Notes of AAAI Fall Symposium on Communicative Action in Humans and Machines, Boston, MA.

Cowan, N. 2001. The Magical Number of 4 in Short-Term Memory: A Reconsideration of Mental Storage Capacity. *Behavioural and Brain Sciences*, **24**(1): 87–185.

Daly-Jones, O., Bevan, N., Thomas, C. 1999. INUSE 6.2 Handbook of User-Centred Design D6.2.1 Version 1.2. 12 February 1999.

Danieli M., Gerbino E. 1995. Metrics for Evaluating Dialogue Strategies in a Spoken Language System. Working Notes of the AAAI Spring Symposium on Empirical Methods in Discourse Interpretation and Generation, pp. 34–9.

DeCarlo, D., Stone, M., Revilla, C. Venditti, J. 2002. Specifying and Animating Facial Signals for Discourse in Embodied Conversational Agents. *Computer Animation and Virtual Worlds*, **15**(1): 27–38.

Dey, AK 2001. Understanding and Using Context. *Personal and Ubiquitous Computing*, **5**: 20–4.

DiEugenio, B., Glass, M. 2004. The Kappa Statistic: A Second Look. *Computational Linguistics*, **30**(1): 95–102.

Dix, A., Finlay, J., Abowd, G. and Beale, R. 1998. *Human Computer Interaction*. 2nd edn. Prentice Hall.

Douglas, C. E., Campbell, N., Cowie, R., Roach, P. 2003. Emotional speech: Towards a new generation of databases. *Speech Communication*, **40**: 33–60.

Dreyfus, H. What Computers Can't Do: *The Limits of Artificial Intelligence*. Harper and Row; Revised edition, 1979.

Duncan, S., Jr., and Fiske, D. W. 1977. *Face-to-face interaction: Research, methods and theory*. Hillsdale, New Jersey: Lawrence Erlbaum. Distributed by John Wiley and Sons.

Dybkjaer, L., Bernsen, N. O., Dybkjaer, H. 1996. Evaluation of Spoken Dialogue Systems. *Proceedings of the 11th Twente Workshop on Language Technology*, Twente.

Echihabi, A., Hermjakob, U., Hovy, E., Marcu, D., Melz, E., Ravichandran, D. 2006. In Strzalkowski, T. and Harabagiu, S. (Eds.) *Advances in Open Domain Question Answering*. Springer, pp. 383–406.

Elman, J. L. 1999. The Emergence of Language: A Conspiracy Theory. In B. MacWhinney (Ed.) *Emergence of Language*. Hillsdale, New Jersey: Lawrence Earlbaum Associates.

Elman, J. L., Bates, E. A., Johnson, M., Karmiloff-Smith, A., Parisi, D., Plunkett, K. 1996. *Rethinking Innateness: A Connectionist Perspective on Development*. Cambridge, Mass.: MIT Press.

Erman, L. D., Hayes-Roth, F., Lesser, V. R., Reddy, D. R. 1980. The HEARSAY-II Speech Understanding System: Integrating Knowledge to Resolve Uncertainty. *Computing Surveys*, Volume 12, Number 2, ACM, pp. 213–53.

Fais, L. 1994. Conversation as Collaboration: Some Syntactic Evidence. *Speech Communication*, **15**(3–4): 231–242.

Feldman, R. S., Rim, B. 1991. *Fundamentals of Nonverbal Behavior*, Cambridge: Cambridge University Press.

Fikes, R., Nilsson, N. 1971. STRIPS: A New Approach to the Application of Theorem Proving to Problem Solving. *Artificial Intelligence*, **2**: 189–203.

Fischer, G. 2001. User Modeling in Human–Computer Interaction. *User Modeling and User-Adapted Interaction*. **11**(1–2): 65–86.

Fraser, N. M. 1997. Spoken Dialogue System Evaluation: A first framework for reporting results. *Proceedings of the 5th European Conference on Speech Communication and Technology*, 1907–1910.

Fraser, N., Gilbert, G. N. 1991. Simulating Speech Systems. *Computer Speech and Language*, **5**: 81–99.

Furui, S. (forthcoming). History and Development of Speech Recognition. In: Chen, F. and Jokinen, K. (Eds.) *New Trends in Speech-based Interactive Systems*. Springer Publishers.

Gallese, V., Fadiga, L., Fogassi, L., Rizzolatti, G. 1996. Action Recognition in the Premotor Cortex. *Brain*, **119**: 593–609.

Galliers, J. R. 1989. A Theoretical Framework for Computer Models of Cooperative Dialogue, Acknowledging Multi-agent Conflict. Tech. rep 17.2, Computer Laboratory, University of Cambridge.

Garfinkel, H. 1967. *Studies in Ethnomethodology*. Englewood Cliffs: Prentice-Hall.

Garrod, S., Doherty, G. 1994. Conversation, Co-ordination and Convention: An empirical Investigation of how Groups Establish Linguistic Conventions. *Cognition*. **53**: 181–215.

Gebhard, P., Kipp, M., Klesen, M., Rist, T. 2003. Authoring Scenes for Adaptive, Interactive Performances. *Proceedings of the second international joint conference on Autonomous agents and multiagent systems, AAMAS '03*. New York: ACM Press, pp. 725–32.

Georgila, K., Henderson, J., Lemon, O. 2005. Learning User Simulations for Information State Update Dialogue Systems. *Proceedings of Interspeech-Eurospeech*, Lisbon, Portugal.

Gibbon, D., Mertins, I., Moore, R. (Eds.) 2000. *Handbook of Multimodal and Spoken Dialogue Systems. Resources, Terminology and Product Evaluation*. Kluwer.

Ginsburg J. 1994. An update semantics for dialogue. In H. Bunt (Ed.) *Proceedings of the International Workshop on Computational Semantics*, pp. 111–120. ITK, Tilburg.

Gmytrasiewicz, P. J., Durfee, E. H. 1993. Elements of Utilitarian Theory of Knowledge and Action. *Proceedings of the Twelfth International Joint Conference on Artificial Intelligence*, pp. 396–402.

Gmytrasiewicz, P. J., Durfee, E. H. 2000. Rational Coordination in Multi-Agent Environments. *Autonomous Agents and Multi-Agent Systems Journal*. **3**(4): 319–50.

Gmytrasiewicz, P. J., Durfee, E. H., Rosenschein, J. S. 1995. *Towards Rational Communicative Behavior*. AAAI Fall Symposium on Embodied Language, Cambridge, Mass.: AAAI Press.

Goffman, E. 1969. *Presentation of Self in Everyday Life*. New York: Anchor.

Goffman, E. 1970. *Strategic Interaction*. Oxford: Blackwell.

Goffman, E. 1974. *Frame Analysis*. New York: Harper & Row.

Goodwin, C. 1981. *Conversational Organization: Interaction between Speakers and Hearers*. New York: Academic Press.

Gorin, A. L., Riccardi, G., Wright, J. H. 1997. How May I Help You? *Speech Communications*, **23**(1/2): 113–27

Grice, H. P. 1957. Meaning. *The Philosophical Review*, **66**(3): 377–88. Reprinted in Grice (1989).

Grice, H. P. 1975. Logic and conversation. In Cole, P. and Morgan, J. L. (Eds.). *Syntax and Semantics*. Vol **3**: Speech Acts, pp. 41–58. New York: Academic Press.

Grice, H. P. 1989. *Studies in the Way of Words*. Cambridge Mass.: Harvard University Press.

Grosz, B. J. 1977. *The Representation and Use of Focus in Dialogue Understanding*. SRI Stanford Research Institute, Stanford, CA.

Grosz, B. J., Hirschberg, J. 1992. Some Intonational Characteristics of Discourse. Proceedings of the ICSLP.

Grosz, B. J., Joshi, A. K., Weinstein, S. 1995. Centering: A Framework for Modeling the Local Coherence of Discourse. *Computational Linguistics*, **21**(2): 203–25.

Grosz, B. J., Kraus, S. 1995. Collaborative Plans for Complex Group Action. Tech. Report TR-20–95, Harvard University, Center for Research in Computing Technology.

Grosz, B. J., Sidner, C. 1990. Plans for Discourse. In P. R. Cohen, J. Morgan, and M. E. Pollack, (Eds.). *Intentions in Communication*. Cambridge, Mass.: MIT Press, pp. 417–44.

Grosz, B. J., Sidner, C. L. 1986. Attention, Intentions, and the Structure of Discourse. *Computational Linguistics*, **12**(3): 175–203.

Gruber, T. R. 1993. A Translation Approach to Portable Ontologies. *Knowledge Acquisition*, **5**(2): 199–220.

Guinn, C. I. 1996. Mechanisms for Mixed-initiative Human-computer Collaborative Discourse. *Proceedings of the 34th Annual Meeting of the Association for Computational Linguistics*, pp. 278–285.

Gumperz, J., Hymes, D. (Eds.). 1972. *Directions in Sociolinguistics: The Ethnography of Communication*. New York: Holt, Rinehart and Winston.

Gustafson, J., Lindberg, N., Lundeberg, M. (1999). The August Spoken Dialog System. *Proceedings of Eurospeech'99*, Budapest, Hungary.

Habermas, J. 1976. What is Universal Pragmatics? The English translation in Habermas (1984) *Communication and the Evolution of Society*. Translated and with an introduction by Thomas McCarthy 1984. Polity Press, Newcastle.

Hahn, U., Strube, M. 1997. Centering-in-the-large: Computing Referential Discourse Segments. *Proceedings of the 35th Annual Meeting of the Association for Computational Linguistics, Madrid*, pp. 104–11.

Hanna, P., O'Neill, I. M., Wootton, C., McTear, M. 2007. Promoting Extension and Reuse within a Spoken Dialogue Manager: An Evaluation of the Queen's Communicator. *ACM Transactions in Spoken Language Processing*, **4**(3).

Harabagiu, S., Pasca, M., Maiorano, S. 2000. Experiments with Open-domain Textual Question Answering. *Proceedings of COLING-2000*, pp. 292–298.

Halliday, M. A. K. 1994. *Introduction to Functional Grammar*. London: Edward Arnold.

Halliday, M. A. K. 1970a. *A Course in Spoken English: Intonation*. Oxford: Oxford University Press.

Halliday, M. A. K. 1970b. Language Structure and Language Function. In: *New Horizons in Linguistics,* John Lyons (Ed.), Harmondsworth, England: Penguin, pp. 140–64.

Halliday, M. A. K. 1969. Categories of the Theory of Grammar. *Word* **17** (3). 241–292.

Halliday, M. A. K. 1967. Notes on Transitivity and Theme in English. *Journal of Linguistics*, **3**(2): 199–244.

Halliday, M. A. K., 1994. *An Introduction to Functional Grammar*. (2nd edition). London: Edward Arnold.

Harnard, S. 1990. The Symbol Grounding Problem. *Physical D*, **42**: 335–46.

Hasida, K, Den, Y., Nagao, K., Kashioka, H., Sakai, K., Shimazu, A. 1995. Dialeague: a proposal of a context for evaluating natural language dialogue systems. *Proceedings of the 1st Annual Meeting of the Japanese Natural Language Processing Society*, pp. 309–312. (in Japanese).

Have, P. 1999. *Doing Conversation Analysis: A Practical Guide*. SAGE Publications.

Heeman, P. A., Allen, J. F. 1997. Intonational boundaries, speech repairs, and discourse markers: Modelling spoken dialog. *Proceedings of the Thirty-Fifth Annual Meeting of the Association for Computational Linguistics and Eighth Conference of the European Chapter of the Association for Computational Linguistics*, pp. 254–261.

Heeman, P. A., Hirst, G. 1995. Collaborating on Referring Expressions. Computational. *Linguistics*, **2**(3): 351–82.

Heritage, J. 1984. A Change-of-state Token and Aspects of its Sequential Placement. In Atkinson, M. and Atkinson, M., Heritage, J. (Eds.) *Structures of Social Action: Studies in Conversation Analysis*. Cambridge: Cambridge University Press, pp. 299–347.

Heritage, J. 1989. Current developments in conversation analysis. In Roger, D., Bull, P. (Eds.) *Conversation: an interdisciplinary perspective*, pp. 21–47. Multilingual Matters, Clevedon.

Hinze, A., Buchanan, G. 2005. Context-awareness in Mobile Tourist Information Systems: Challenges for User Interaction. In *Proceedings of the Workshop on Context in Mobile HCI, in conjunction with Mobile HCI*,Salzburg, Austria.

Hirasawa, J., Nakano, M., Kawabata, T., Aikawa, K. 1999. Effects of System Barge-in Responses on User Impressions. *The Sixth European Conference on Speech Communication and Technology*, **3**: 1391–4.

Hirschberg, J., Litman, D. 1993. Empirical Studies on the Disambiguation of Cue Phrases. *Computational Linguistics*, **19**(3): 501–30.

Hirschberg, J., Nakatani, C. H. 1996. A Prosodic Analysis of Discourse Segments in Direction-Giving Monologues. *Proceedings of the Annual Meeting of the Association for Computational Linguistics*, pp. 286–93.

Hirschberg, J., Nakatani, C. 1998. Acoustic Indicators of Topic Segmentation. *Proceedings of the International Conference on Spoken Language Processing*, pp. 976–979, Sydney, Australia.

Hobbs, J. 1979. Coherence and coreference. *Cognitive Science*, **3**(1): 67–90.

Horvitz, E. 1987. Reasoning about Beliefs and Actions under Computational Resource Constraints, Third Workshop on Uncertainty in Artificial Intelligence, Seattle, Washington. July 1987. Association for Uncertainty and Artificial Intelligence. pp. 429–444. Also in Kanal, L., Levitt, T. and Lemmer, J. (Eds.) *Uncertainty in Artificial Intelligence 3*, Amsterdam: Elsevier/North-Holland. pp. 301–324.

Horvitz, E. J. 1988. Reasoning Under Varying and Uncertain Resource Constraints. *Proceedings of the 7th National Conference on Artificial Intelligence*, Minneapolis, MN. Morgan Kaufmann, San Mateo, CA. pp. 111–16.

Hovy, E. H. 1988. *Generating Natural Language under Pragmatic Constraints*. Lawrence Erlbaum Associates, Hillsdale, NJ.

Hurtig, T., Jokinen, K. 2006. Modality Fusion in a Route Navigation System. *Proceedings of the IUI 2006 Workshop on Effective Multimodal Dialogue Interfaces*, pp. 19–24.

Hurtig, T., Jokinen, K. 2005. On Multimodal Route Navigation in PDAs. *Proceedings of the 2nd Baltic Conference on Human Language Technologies*, Tallinn, Estonia, pp. 261–6.

Hymes, D. 1971. Sociolinguistics and Ethnography of Speaking. In E. Ardener (Ed.) *Social Anthropology and Linguistics. Association of Social Anthropologists*. Monograph 10. London: Tavistock, pp. 47–93.

Ichikawa, A., Araki, M., Ishizaki, M., Itabashi, S., Itoh, T., Kashioka, H., Kato, K., Kikuchi, H., Kumagai, T., Kurematsu, A., Koiso, H., Tamoto, M., Tutiya, S., Nakazato, S., Horiuchi, Y., Maekawa, K., Yamashita, Y., Yoshimura, T. 1998. Standardising Annotation Schemes for Japanese Discourse. *Proceedings of the First International Conference on Language Resources and Evaluation*, pp. 731–6, Granada, Spain.

Isard, A., McKelvie, D., Cappelli, B., Dybkjær, L., Evert, S., Fitschen, A., Heid, U., Kipp, M., Klein, M., Mengel, A., Møller, M. B. Reithinger, N. 1998. Specification of Workbench Architecture. MATE Deliverable D3.1.

Isbister, K., Nakanishi, H., Ishida, T., Nass, C. 2000. Helper Agent: Designing an Assistant for Human-human Interaction in a Virtual Meeting Space. *Proceedings of the SIGCHI conference on Human factors in computing systems (CHI'00)*, New York, NY, USA, ACM Press, pp. 57–64.

Ishiguro, H. 2005. Android science: Toward a new Cross-Interdisciplinary Framework. *Proceedings of the CogSci 2005 Workshop Toward Social Mechanisms of Android Science*. pp. 1–6.

Janin, A., Baron, D., Edwards, J. Ellis, D., Gelbart, D., Morgan, N. Peskin, B., Pfau, T., Shriberg, E., Stolcke, A., Wooters. C. 2003. The ICSI meeting corpus. *Proceedings of IEEE International Conference on Acoustics, Speech and Signal Processing (ICASSP)*, Hong-Kong, pp. 364–7.

Jekat, S., Klein, A., Maier, E., Maleck, I., Mast, M., Quantz, J. 1995. Dialogue Acts in VERBMO-BIL. Tech. Rep. 65, BMBF Verbmobil Report.

Jokinen, K. 1995. Rationality in Constructive Dialogue Management. *Proceedings of the AAAI-95 Fall Symposium Series. Rational Agency: Concepts, Theories, Models & Applications*. MIT, Boston. pp. 89–93.

Jokinen, K. 1996a. Reasoning about Coherent and Cooperative System Responses. In G. Adorni and M. Zock (Eds.) *Trends in Natural Language Generation. An Artificial Intelligence Perspective*. Berlin: Springer-Verlag, pp. 168–87.

Jokinen, K. 1996b. Goal Formulation Based on Communicative Principles. *Proceedings of The 16th International Conference on Computational Linguistics (COLING-96)*, Copenhagen, Denmark. pp. 598–603.

Jokinen, K. 1996c. Cooperative Response Planning in CDM. *Proceedings of the 11th Twente Workshop on Language Technology: Dialogue Management in Natural Language Processing Systems*, pp. 159–68. The Netherlands.

Jokinen, K. 2000. Learning Dialogue Systems. In L. Dybkjaer (Ed.) *LREC 2000 Workshop: From Spoken Dialogue to Full Natural Interactive Dialogue – Theory*, Empirical Analysis and Evaluation, Athens, pp. 13–17.

Jokinen, K. 2003. Natural Interaction in Spoken Dialogue Systems. In Stephanidis, C. and Jacko, J. (Eds.) Human-Computer Interaction: Theory and Practice (Part II), volume 4, pp. 730–734, Mahwah, New Jersey. Lawrence Eribaum Associates.

Jokinen, K. 2004 Communicative Competence and Adaptation in a Spoken Dialogue System. *Proceedings of the 8th International Conference on Spoken Language Processing (ICSLP-Interspeech)*, pp. 1729–32, Jeju, Korea.

Jokinen, K. 2005a. Challenges for Adaptive Conversational Agents. *Proceedings of the 2nd Baltic Conference on Human Language Technologies*, Tallinn, pp. 51–60.

Jokinen, K. 2005b. Adaptation and User Expertise Modelling in AthosMail. Universal Access in the Information Society, pp. 1–19. http://www.springerlink.com/(nsj13c453g4e5ceioloej0ii)/app/home/contribution.asp?referrer = parent&backto = issue,1,10;journal,1,15;linkingpublicationresults,1:107725,1.

Jokinen, K. 2005c. Finnish Discourse Syntax Grammar. In A. Arppe et al. (Eds.) *Inquiries into Words, Constraints and Contexts*. Festschrift for Kimmo Koskenniemi on his 60th Birthday. CSLI Studies in Computational Linguistics ONLINE (Series Editor Ann Copestake). ISSN 1557–5772. CSLI Publications, Stanford, California, pp. 227–41. Available at http://cslipublications.stanford.edu/site/SCLO.html.

Jokinen, K. 2008. User Interaction in Mobile Navigation Applications. In L. Meng, A. Zipf and S. Winter (Eds.) *Map-based Mobile Services – Design, Interaction and Usability*, pp. 168–97. Springer Series on Geoinformatics.

Jokinen, K., Gambäck, B. 2004. DUMAS – Adaptation and Robust Information Processing for Mobile Speech Interfaces. *Proceedings of The 1st Baltic Conference "Human Language Technologies – The Baltic Perspective"*, pp. 115–20, Riga, Latvia.

Jokinen, K., Hurtig, T. 2006. User Expectations and Real Experience on a Multimodal Interactive System. *Proceedings of Interspeech-2006*, Pittsburgh, US.

Jokinen, K., Hurtig, T., Hynnä, K., Kanto, K., Kerminen, A., Kaipainen, M. 2001. Self-organizing Dialogue Management. In Isahara, H. and Ma, Q. (Eds.) NLPRS2001 *Proceedings of the 2nd Workshop on Natural Language Processing and Neural Networks*, pp. 77–84, Tokyo, Japan.

Jokinen, K., Kanto, K. 2004. User Expertise Modelling and Adaptivity in a Speech-based E-mail System. *Proceedings of the 42nd Annual Meeting of the Association of Computational Linguistics, ACL 2004*, Barcelona, Spain.

Jokinen, K., Kanto, K., Kerminen, A., J. Rissanen 2004. Evaluation of Adaptivity and User Expertise in a Speech-based E-mail System. *Proceedings of the COLING Satellite Workshop Robust and Adaptive Information Processing for Mobile Speech Interfaces*, Geneva, Switzerland, pp. 44–52.

Jokinen, K., A. Kerminen, M. Kaipainen, T. Jauhiainen, G. Wilcock, M. Turunen, J. Hakulinen, J. Kuusisto, K. Lagus 2002. Adaptive Dialogue Systems: Interaction with Interact. In Jokinen, K and S. McRoy (Eds.) *Proceedings of the 3rd SIGDial Workshop on Discourse and Dialogue*, Philadelphia, USA, pp. 64–73.

Jokinen, K., McTear, M. (forthcoming) Spoken Dialogue Systems. *To appear in the Synthesis Lectures on Human Language Technologies*. Edited by Graeme Hirst. Morgan and Claypool Publishers.

Jokinen, K., Paggio, P., Navarretta, C. 2008. Distinguishing the Communicative Functions of Gestures – An Experiment with Annotated Gesture data. Machine-Learning in Multimodal Interaction. *Springer Lecture Notes on Computer Science*, Springer, pp. 38–49.

Jokinen, K., Raike A. 2003. Multimodality – Technology, Visions and Demands for the Future. *Proceedings of the 1st Nordic Symposium on Multimodal Interfaces*, Copenhagen, Denmark, pp. 239–251.

Jokinen, K., Rissanen, J., Keränen, H., Kanto, K. 2002. Learning Interaction Patterns for Adaptive User Interfaces. *Adjunct Proceedings of the 7th ERCIM Workshop User Interfaces for all*. Paris, pp. 53–58.

Jokinen, K., Tanaka, H., Yokoo, A. 1998. Context Management with Topics for Spoken Language Systems. *Proceedings of the Joint International Conference of Computational Linguistics and the Association for Computational Linguistics (COLING-ACL'98)*. Montreal. pp. 631–7.

Jokinen, K., G. Wilcock 2003. Adaptivity and Response Generation in a Spoken Dialogue System. In van Kuppevelt, J. and R. W. Smith (Eds.) *Current and New Directions in Discourse and Dialogue*. Kluwer Academic Publishers, pp. 213–34.

Jokinen, K., G. Wilcock 2006. Contextual Inferences in Intercultural Communication. In M. Suominen et al. (Eds.), *A Man of Measure: Festschrift in Honour of Fred Karlsson. Special Supplement to SKY Journal of Linguistics*, Vol. **19**. The Linguistic Association of Finland, Turku, pp. 291–300. Available also online at: http://www.ling.helsinki.fi/sky/julkaisut/sky2006special.shtml

Joshi, A., Webber, B., Sag, I. (Eds.) 1981. *Elements of Discourse Understanding*, Cambridge, Mass.: Cambridge University Press.

Joshi, A., Webber, B. L., Weischedel, R. M. 1984. Preventing False Inferences. *Proceedings of the 10th COLING*, pp. 34–138.

Jurafsky, D., Shriberg, E., Biasea, D. 1997. Switchboard SWBD-DAMSL, Shallow-Discourse-Function Annotation; Coders Manual. Tech. Rep. 97-02, University of Colorado Institute of Cognitive Science. Draft 13.

Jurafsky, D., Shriberg, E., Fox, B., Curl, T. 1998. Lexical, Prosodic, and Syntactic cues for Dialog Acts. *Proceedings of the ACL/COLING-98 Workshop on Discourse Relations and Discourse Markers*, pp. 114–120.

Jönsson, A. 1997. A Model for Habitable and Efficient Dialogue Management for Natural Language Interaction. *Natural Language Engineering* **3**(2–3): 103–22.

Kaiser, E., Trueswell, J. C. 2004. The Role of Discourse Context in the Processing of a Flexible Word-Order Language. *Cognition* **94**(2): 113–47.

Kanto, K., Cheadle, M., Gambäck, B., Hansen, P., Jokinen, K., Keränen, H., Rissanen, J. 2003. Multi-session Group Scenarios for Speech Interface Design. In C. Stephanidis and J. Jacko (Eds.) *Human-Computer Interaction: Theory and Practice (Part II)*, Volume 2, pp. 676–80, Mahwah, New Jersey, June. Lawrence Erlbaum Associates.

Karttunen, L. 1971. Discourse Referents. In J. McCawley (Ed.) *Syntax and Semantics 7: Notes from the Linguistic Underground*. New York: Academic Press, pp. 363–86.

Kaski, S. 1998. Dimensionality Reduction by Random Mapping: Fast Similarity Computation for Clustering. *Proceedings of IJCNN'98*, volume 1, pp. 413–418. Piscataway, NJ: IEEE Service Center.

Katagiri, Y. 2005. Interactional Alignment in Collaborative Problem Solving Dialogues, *Proceedings of the 9th International Pragmatics Conference*, Riva del Garda Italy.

Kearns, M., Isbell, C., Singh, S., Litman, D., Howe, J. 2002. CobotDS: A Spoken Dialogue System for Chat. In *Proceedings of the Eighteenth National COnference on Artificial Intelligence (AAAI)*, pp. 425–430.

Keizer, S., Akker, R. op den, Nijholt, A. 2002. Dialogue Act Recognition with Bayesian Network for Dutch Dialogues. In Jokinen, K., McRoy, S. (Eds.) *Proceedings of the 3rd SIGDial Workshop on Discourse and Dialogue*, Philadelphia, US, pp. 88–94.

Kendon, A. 2004. *Gesture: Visible action as utterance*. Cambridge: Cambridge University Press.

Kerminen, A. and Jokinen K. 2003. Distributed Dialogue Management. In Jokinen, K., Gambäck B., Black, W. J., Catizone, R., Wilks, Y. (Eds.) *Proceedings of the EACL Workshop on Dialogue Systems: Interaction, Adaptation and Styles of Management*. Budapest, Hungary, pp. 55–66.

Kipp, M. 1998. The Neural Path to Dialogue Acts. Proceedings of the 13th European Conference on Artificial Intelligence (ECAI). pp. 175–179.

Kipp, M. 2001. Anvil – A Generic Annotation Tool for Multimodal Dialogue. *Proceedings of the Seventh European Conference on Speech Communication and Technology*, pp. 1367–70.

Kita, S. 2003. Interplay of Gaze, Hand, Torso Orientation, and Language in Pointing. In Kita, S. (Ed.) *Pointing. Where Language, Culture, and Cognition Meet*. London: Lawrence Erlbaum Associates, pp. 307–28.

Kita, K., Fukui, Y., Nagata, M., Morimoto, T. 1996. Automatic Acquisition of Probabilistic Dialogue Models. *Proceedings of the 4th International Conference on Spoken Language Processing*, pp. 196–9.

Kobsa, A., Wahlster, W. 1989. *User Models in Daialog Systems.* Springer, Berlin, Heidelberg.

Koeller, A., Kruijff, G.-J. 2004. Talking Robots with LEGO MindStorms. *Proceedings of the 20th COLING*, Geneva.

Koiso, H., Horiuchi, Y., Tutiya, S., Ichikawa, A., Den, Y. 1998. An Analysis of Turn taking and Backchannels Based on Prosodic and Syntactic Features in Japanese Map Task dialogs. *Language and Speech*, **41**(3–4): 295–321.

Krahmer, E., Swerts, M. 2006. Testing the Effect of Audiovisual Cues to Prominence via a Reaction-time Experiment, *Proceedings of the International Conference on Spoken Language Processing (Interspeech 2006)*, Pittsburgh, PA, USA.

Krahmer, E. J., Swerts, M. 2007. Perceiving Focus. In Lee, C., Gordon, M., Buring, D. (Eds.), *Topic and Focus: a Cross-linguistic perspectives on meaning and intonation.* Studies in Linguistics and Philosophy, 82. Dordrecht: Springer, pp. 121–137.

Krahmer, E., Swerts, M., Theune, M., Weegels, M. 1999. Problem Spotting in Human-Machine Interaction. *Proceedings of Eurospeech '99*, pp. 1423–1426. Budapest, Hungary.

Kray, C., Laakso, K., Elting, C., Coors, V. 2003. Presenting Route Instructions on Mobile Devices. In *Proceedings of IUI 03*, pp. 117–124. Miami Beach, FL, ACM Press.

Kruijff, G-J., Zender, H., Jensfelt, P. and Christensen, H. I. 2007. Situated Dialogue and Spatial Organization: What, Where ... and Why? *International Journal of Advanced Robotic Systems, Special Issue on Human and Robot Interactive Communication.* Vol. **4**, No. 2.

Kruijff-Korbayová, I. and Steedman, M. (Eds.) Discourse and Information Structure. *Special Issue of the Journal of Logic, Language and Information*, **12**(3): 249–259

Kudo, T., Matsumoto, Y. 2001. Chunking with Support Vector Machines. *Proceedings of the Second Conference on North American Chapter of the Association for Computational Linguistics (NAACL-01)*, pp. 192–199.

Kuppevelt, J. van 1995. Discourse Structure, Topicality and Questioning. *Linguistics*, **31**: 109–147.

Kurzweil, R. 1999. *The Age of Spiritual Machines.* New York: Penguin Books.

Lady Lovelace http://www.abelard.org/turpap/turpap.htm.

Larsson, S., Berman, A., Bos, J., Grönqvist, L., Ljunglöf, P., Traum, D. 2000. TrindiKit 2.0 Manual.

Larsson, S., Cooper, R., Ericsson, S. 2001. menu2dialog. In Jokinen, K. (Ed.): *Knowledge and Reasoning in Practical Dialogue Systems, Workshop Program, IJCAI-2001.* pp. 41–5.

Larsson, S., Traum, D. 2000. Information State and Dialogue Management in the TRINDI Dialogue Move Engine Toolkit. *Natural Language Engineering*, **6**(3–4): 323–40.

Laurel, B. 1993. *Computers as Theatre.* Boston, Mass.: Addison Wesley.

Lauria, S., Bugmann, G., Kyriacou, T., Bos, J., Klein, E. 2001. Training Personal Robots using Natural Language Instruction. *IEEE Intelligent Systems*, **16**(3): 38–45.

Leech, G. N. 1983. *Principles of Pragmatics.* London: Longman.

Lemon, O., Bracy, A., Gruenstein, A., Peters, S. 2001 Information States in a Multi-modal Dialogue System for Human-Robot Conversation. *Proceedings of Bi-Dialog, the 5th Workshop on Formal Semantics and Pragmatics of Dialogue*, pp. 57–67.

Lenat, Douglas B., Guha, R. V. 1990. *Building Large Knowledge-Based Systems.* Reading, Mass.: Addison-Wesley.

Lendevai, P., Bosch, A. van den, Krahmer, E. 2003. Machine Learning for Shallow Interpretation of User Utterances in Spoken Dialogue Systems. In Jokinen, K., Gambäck B., Black, W. J., Catizone, R., Wilks, Y. (Eds.) *Proceedings of the EACL Workshop on Dialogue Systems: Interaction, Adaptation and Styles of Management*, pp. 71–80, Budapest, Hungary.

Lesh, N. Rich, C., Sidner, C. L. 1998. Using Plan Recognition in Human-Computer Collaboration. Merl Technical Report.

Levelt, W. J. M. 1989. *Speaking: From Intention to Articulation*. Cambridge, MA: The MIT Press.

Levesque, H. J., Cohen, P. R., Nunes, J. H. T. 1990. On acting together. *Proceedings of AAAI-90*, pp. 94–99. Boston, Massachusetts.

Levin, E., Pieraccini, R. 1997. A Stochastic Model of Computer-human Interaction for Learning Dialogue Strategies. *Proceedings of Eurospeech*, pp. 1883–6, Rhodes, Greece.

Levin, E., Pieraccini, R., Eckert, W. 2000. A Stochastic Model of Human-machine Interaction for Learning Dialog Strategies, *IEEE Transactions on speech and audio processing*, **8**(1): 11–23.

Levinson, S. 1979. Activity Types and Language. *Linguistics*, **17**(5/6): 365–399, Mouton, The Hague.

Levinson, S. C. 1981. Some Pre-observations on the Modelling of Dialogue. *Discourse Processes*, **4**(2): 93–116.

Levinson, S. 1983. *Pragmatics*. Cambridge: Cambridge University Press.

Levinson S. C. 1987. Minimization and Conversational Inference. In M. Papi and J. Verschueren (Eds.), *The Pragmatic Perspective: Proceedings of the International Conference on Pragmatics at Viareggio*, pp. 61–129. Amsterdam: J. Benjamins.

Levy, D., Batacharia, B., Catizone, R., Krotov, A., Wilks, Y. 1999. CONVERSE – a Conversational Companion. In Wilks, Y. (Ed.) *Machine Conversations*, pp. 205–16. Amsterdam: Kluwer Academic Publishers.

Lin, J., Quan, D. Sinha, V., Bakshi, K., Huynh, D., Katz, B., and Karger, D. R. 2003. The role of context in question answering systems. *In CHI '03 Extended Abstracts on Human Factors in Computing Systems. CHI '03. ACM, New York, NY*, 1006–1007

Litman, D. J. 1985. Plan Recognition and Discourse Analysis: in Integrated Approach for Understanding Dialogues. Tech. Report. 170, Department of Computer Science, University of Rochester.

Litman, D. J., Allen, J. F. 1984. A Plan Recognition Model for Clarification Subdialogues. *Proceedings of the 10th international Conference on Computational Linguistics and 22nd Annual Meeting on Association For Computational Linguistics*, pp. 302–11.

Litman, D. J., Allen, J. 1987. A Plan Recognition Model for Subdialogues in Conversation. *Cognitive Science*, **11**(2): 163–200.

Litman, D. J., Pan. S. 2002. Designing and Evaluating an Adaptive Spoken Dialogue System. *User Modeling and User-Adapted Interaction*, **12**(2–3): 111–37.

Litman, D., Kearns, M., Singh, S., Walker, M. 2000. Automatic Optimization of Dialogue Management. *Proceedings of the 18th COLING*, pp. 502–8.

Lochbaum, K. E. 1994. Using Collaborative Plans to Model the Intentional Structure of Discourse. Tech. Report TR-25–94, Harvard University, Center for Research in Computing Technology.

Loecher, M. S. Schoeder, C. and Thomas, C. W. 2000. *Proteus: insight from 2020*. The Copernicus Institute Press.

Lopez Cozar, R. and Araki, M. 2005. *Spoken, Multilingual and Multimodal Dialogue Systems*. Chichester: John Wiley and Sons Ltd.

López-Cózar, R., de la Torre, A., Segura, J., and Rubio, A. 2003. Assessment of Dialogue Systems by Means of a New Simulation Technique. *Speech Communication*, **40**: 387–407.

Luff, P., Gilbert, N. G. and Frohlich, F. 1990. *Computers and Conversation*. London: Academic Press.

Luhman, N. 1979. *Trust*. In *Trust and Power: Two Works by Niklas Luhman*. Chichester: John Wiley and Sons Ltd.

Lund, K., Burgess, C. 1996. Producing High-dimensional Semantic Spaces from Lexical Co-occurrence. Behavior Research Methods, Instruments & Computers, **28**(2): 203–208.

Maes, P. (Ed.) 1990. *Designing Autonomous Agents: Theory and practice from Biology to Engineering and Back*. Cambridge, Mass.: MIT Press.

Malinowski, B. 1923. The Problem of Meaning in Primitive Languages. In Ogden, C. and Richards, I. (Eds.) *The Meaning of Meaning*. London: Routledge and Kegan Paul, pp. 146–52.

Mandler, J. 2004. *The Foundations of Mind: Origins of conceptual thought*. Oxford: Oxford University Press.

Martin, D., Cheyer, A., Moran, D. 1998. Building Distributed Software Systems with the Open Agent Architecture. *Proceedings of the 3rd International Conference on the Practical Application of Intelligent Agents and Multi-Agent Technology*, Blackpool, UK. The Practical Application Company, Ltd.

Martinovski, B. 2004. Communication as Reproduction of Self vs. Ethics of Otherness. Nordic Conference on Intercultural Communication, Göteborg University, Göteborg, Sweden.

Mavridis, N., Roy, D. 2006. Grounded Situation Models for Robots: Where Words and Percepts Meet. IEEE/RSJ International Conference on Intelligent Robots and Systems (IROS).

Maybury, M., Wahlster, W. 1998. *Readings in Intelligent User Interfaces*. Los Altos, California: Morgan Kaufmann.

McCarthy, J. 1979 Ascribing Mental Qualities to Machines. In Ringle, M. (Ed.) *Philosophical Perspectives in Artificial Intelligence*. Atlantic Highlands, New Jersey: Humanities Press. Reprinted in McCarthy, J. 1990. *Formalizing Common Sense*. Edited by V. Lifschitz. Ablex.

McCoy, K. F. 1988. Reasoning on a Highlighted User Model to Respond to Misconceptions. *Computational Linguistics*, **14**(3): 52–63.

McCoy, K., Cheng, J. 1990. Focus of Attention: Constraining what Can Be Said Next. In Paris, C. L., Swartout, W. R., Moore, J. C. (Eds.) *Natural Language Generation in Artificial Intelligence and Computational Linguistics*, pp. 103–24, Norwell, Mass.: Kluwer Academic Publishers.

McGlashan, S., Fraser, N., Gilbert, N., Bilange, E., Heisterkamp, P., Youd, N. 1992. Dialogue Management for Telephone Information Services Systems. *In Proceedings of the Third Conference on Applied Natural Language Processing, Trento, Italy,* pp. 245–246.

McGuinness, D. L. 2002. Ontologies Come of Age. In Fensel, D., Hendler, J., Lieberman, H. and Wahlster, W. (Eds.) *Spinning the Semantic Web: Bringing the world wide web to its full potential*. Cambridge, Mass.: MIT Press.

McKeown, K. 1985. *Text Generation: Using Discourse Strategies and Focus Constraints to Generate Natural Language Text*. Cambridge: Cambridge University Press, Cambridge.

McNeill, D. 1992. *Hand and Mind: What Gestures Reveal about Thought*. Chicago: University of Chicago Press.

McRoy, S. W., Hirst, G. 1995. The Repair of Speech Act Misunderstandings by Abductive Inference. *Computational Linguistics*, **21**(4): 5–478.

McTear, M. 2004. *Spoken Dialogue Technology: toward the conversational user interface*. London: Springer Verlag.

McTear, M. F. 2006. Handling Miscommunication in Spoken Dialogue Systems Why Bother? In Dybkjaer, L. and Minker, W. (Eds.) *Recent Trends in Discourse and Dialogue*, Springer.

Miikkulainen, R. 1993. *Sub-symbolic Natural Language Processing: An Integrated Model of Scripts, Lexicon, and Memory*. MIT Press, Cambridge.

Miller, G. A. 1956. The Magical Number Seven, Plus or Minus Two: Some Limits on our Capacity for Processing Information. *Psychological Review*, **63**: 81–97.

Milekic, S. 2002. Towards Tangible Virtualities: Tangialities. In Museums and the Web 2002: Selected Papers from an International Conference, Bearman, D. and Trant, J. (Eds.) On-line available at: http://www.archimuse.com/mw2002/papers/milekic/milekic.html.

Minsky, M. 1974. A Framework for Representing Knowledge. MIT-AI Laboratory Memo 306. Shorter version re-printed in Haugeland, J. (Ed.) 1981. *Mind Design*, Cambridge, Mass.: MIT Press, and in Collins, A. and Smith, E. E. (Eds.) 1992. *Cognitive Science*, Morgan-Kaufmann.

Moeschler, J. 1989. *Modélisation du Dialogue*. Paris: Hermès.

Moore, J. D., Paris, C. L. 1993. Planning Text for Advisory Dialogues: Capturing Intentional and Rhetorical Information. *Computational Linguistics*, **19**(4): 651–94.

Moore, J. D., Swartout, W. J. 1992. A Reactive Approach to Explanation: Taking the User's Feedback into Account. In Paris, C. L., Swartout, W. R., and Moore, W. C. (Eds.) *Natural Language Generation in Artificial Intelligence and Computational Linguistics*, pp. 3–48. Norwell, Mass.: Kluwer Academic Publishers.

Mueller, C. Großmann-Hutter, B., Jameson, A., Rummer, R., Wittig, F. 2001. Recognizing Time Pressure and Cognitive Load on the Basis of Speech: An experimental study. In Bauer, M., Vassileva, J., Gmytrasiewicz, P. (Eds.) *Proceedings of the 8th International Conference on User Modeling (UM2000)*, pp. 24–33. Springer, Berlin.

Möller, S. A 2002. New Taxonomy for the Quality of Telephone Services Based on Spoken Dialogue Systems. In Jokinen, K. and S. McRoy (Eds.) *Proceedings of the 3rd SIGDial workshop on Discourse and Dialogue*. Philadelphia, US, 2002, pp. 142–153.

Möller, S. 2005. Quality of Telephone-Based Spoken Dialogue Systems. NewYork: Springer.

Nagata, M., Morimoto, T. 1994. An Information-theoretic Model of Discourse for Next Utterance Type Prediction. *Transactions of Information Processing Society of Japan*, **35**(6): 1050–61.

Nah, F. 2004. A Study on Tolerable Waiting Time: How Long Are Web Users Willing to Wait?, *Behaviour and Information Technology*, **23**(3), pp. 153–163.

Nakatani, C., Hirschberg, J. 1993. A Speech-first Model for Repair Detection and Correction. *Proceedings of the 31st Annual Meeting on Association For Computational Linguistics*, pp. 46–53, Ohio, U.S.

Nakatani, C., Hirschberg, J., Grosz, B. 1995. Discourse Structure in Spoken Language: Studies on Speech Corpora. Working Notes of the AAAI-95 Spring Symposium on Empirical Methods in Discourse Interpretation, Palo Alto.

Nakano, M., Miyazaki, N., Hirasawa, J., Dohsaka, K., Kawabata, T. 1999. Understanding Unsegmented User Utterances in Real-time Spoken Dialogue Systems. *Proceedings of the 37th Annual Meeting of the Association For Computational Linguistics on Computational Linguistics*, pp. 200–7, Maryland, US.

Nakano, M., Miyazaki, N., Yasuda, N., Sugiyama, A., Hirasawa, J., Dohsaka, K., Aikawa, K. 2000. WIT: toolkit for building robust and real-time spoken dialogue systems. In Dybkjær, L., Hasida, K., Traum, D., (Eds.) *Proceedings of the 1st SIGDial workshop on Discourse and Dialogue* – Volume 10, pp. 150–9, Hong Kong.

Nass, C. S. Brave 2005. *Wired for Speech: how voice activates and advances the human-computer relationship*. Cambridge, Mass.: MIT Press,

Neal, J. G., Shapiro, S. C. 1991. Intelligent Multi-media Interface Technology. In Sullivan, J. W., and Tyler, S. W. (Eds.) *Intelligent User Interfaces*, Frontier Series. New York: ACM Press, pp. 11–43.

Negroponte, N. 1979. *The Architecture Machine: Towards a More Human Environment*. Cambridge, Mass.: MIT Press.

Newell, A. Simon, H. 1976. Computer Science as Empirical Inquiry: Symbols and Search. *Communications of the ACM*, **19**: 113–26.

Nielsen, J. 1993. Response Times: The Three Important Limits. In Nielsen 1993. Usability Engineering, Chapter 5,Morgan-Kaufmann.

Nielsen, J. 1994. Heuristic Evaluation. In Nielsen, J., Mack, R. L. (Eds.) *Usability Inspection Methods*, chapter 2, New York: John Wiley and Sons, Inc.

Nivre, J. 1992. Situations, Meaning, and Communication – a Situation Theoretic Approach to Meaning in Language and Communication. PhD Thesis, University of Göteborg.

Norman, D. A. 2004. *Emotional Design: why we Love (or hate) everyday things*. Cambridge, Mass: Basic Books.

Norman, D. A. 1988. *The Psychology of Everyday Things*. New York: Basic Books.

Norman, D., Draper S. (Eds.) 1986 *User Centered System Design: New perspectives on human-computer interaction*. Hillsdale, NJ: Lawrence Erlbaum Associates.

Norros, L., Kaasinen, E., Plomp, J., Rämä, P. (2003) Human–technology Interaction Research and Design. VTT Roadmaps. *VTT Research Notes 2220*. Espoo: VTT Industrial Systems.

O'Neill, I., Hanna, P., Liu, X., McTear, M. 2004. The Queen's Agents: Using Collaborating Object-based Dialogue Agents in the Queen's Communicator. *Proceedings of the 20th International Conference on Computational Linguistics (COLING-2004)* Geneva, Switzerland, pp. 127–133.

Ogden, C. K. Richards, I. A. 1923. *The Meaning of Meaning*. 8th edn, New York: Harcourt, Brace & World, Inc.

Oviatt, S. L. 1996. User-centered Modeling for Spoken Language and Multimodal Interfaces. *IEEE Multimedia*, **3**(4): 26–35.

Oviatt, S., Cohen, P. R., Wu, L., Vergo, J., Duncan, L., Suhm, B., Bers, J., Holzman, T., Winograd, T., Landay, J., Larson, J., Ferro, D. 2000. Designing the User Interface for Multimodal Speech and Pen-based Gesture Applications: State-of-the-Art Systems and Future Research Directions. *Human Computer Interaction*, **15**(4): 263–322.

Paek, T., Horvitz, E. 2000. Conversation as Action under Uncertainty. *Proceedings of 6th Conference on Uncertainty in Artificial Intelligence*, San Francisco, USA. Morgan Kaufmann Publishers, pp. 455–464.

Paiva, A., Dias, J., Sobral, D., Aylett, R., Sobreperez, P., Woods, S., Zoll, C., Hall, L. 2004. Caring for Agents and Agents that Care: Building Empathic Relations with Synthetic Agents. *Proceedings of the Third International Joint Conference on Autonomous Agents and Multiagent Systems (AAMAS '04)*. Washington, DC, USA, IEEE Computer Society, pp. 194–201.

Paris, C. 1988. Tailoring Descriptions to a User's Level of Expertise. *Journal of Computational Linguistics*, **14**(3): 64–78.

Passonneau, R. J., Litman, D. J. 1993. Intention-based Segmentation: human reliability and correlation with linguistic cues. *Proceedings of the ACL*, pp. 148–55.

Peirce, C. S. 1960. *Collected Papers of Charles Sanders Peirce, Volume 2*. Hartshorne, C. and P. Weiss (Eds.), Cambridge, Mass.: Harvard University Press.

Pelachaud, C., Badler, N., Steedman, M. 1996. Generating Facial Expressions for Speech, *Cognitive Science*, **20**: 1–46.

Pelachaud, C., Carofiglio, V., Carolis, B. D., de Rosis, F., Poggi, I. 2002. Embodied contextual agent in information delivering application. *Proceedings of the First international joint conference on Autonomous agents and multiagent systems (AAMAS'02)*, New York: ACM Press, pp. 758–65.

Pelachaud, C., Poggi, I. 2002a. Multimodal Embodied Agents. *The Knowledge Engineering Review*, **17**: 181–196.

Pelachaud, C., I. Poggi 2002b. Subtleties of Facial Expressions in Embodied Agents, *Journal of Visualization and Computer Animation*, **13**: 301–12.

Pickering, M. S. Garrod 2004. Towards a Mechanistic Psychology of Dialogue, *Behavourial and Brain Sciences*, **27**: 169–226.

Pieraccini, R., Levin, E., Eckert, W. 1997. AMICA: the AT&T mixed initiative conversational architecture. *Proceedings of the 5th European Conference on Speech Communicationand Technology*.

Pieraccini, R. (forthcoming). The Industry of Spoken Dialog Systems and the Third Generation of Interactive Applications. In: Chen, F. and Jokinen, K. (Eds.) *New Trends in Speech-based Interactive Systems*. Springer Publishers.

Pignataro, V. 1988. A Computational Approach to Topic and Focus in a Production Model. *Proceedings of COLING*, pp. 515–17.

Plomp, J., Ahola, J., Alahuhta, P., Kaasinen, E., Korhonen, I., Laikari, A., Lappalainen, V., Pakanen, J., Rentto, K., Virtanen, A. 2002. Smart Human Environments. In Sipilä, M. (Ed.). *Communications Technologies. The VTT Roadmaps*. Espoo: Technical Research Centre of Finland, VTT Research Notes 2146, pp. 61–81.

Polifroni, J. Seneff, S., Glass, J., Hazen, T. J. 1998. Evaluation Methodology for a Telephone-based Conversational System. *Proceedings of the 1st International Conference on Language Resources and Evaluation*, pp. 42–50, Granada, Spain.

Power, R. 1979. Organization of Purposeful Dialogue. *Linguistics* **17**: 107–52.

Prendinger, H., Ishizuka, M. 2004. *Life-like Characters: tools, affective functions and applications*. Berlin: Springer.

Price, P., Hirschman, L., Shriberg, E., Wade, E. 1992. Subject-based Evaluation Measures for Interactive Spoken Language Systems. *Proceedings of the DARPA Speech and Natural Language Workshop*, pp. 34–39.

Prince, E. F. 1981. Toward a Taxonomy of Given-new Information. In Cole, P. (Ed.) *Radical Pragmatics*. New York: Academic Press, pp. 223–55.

Rasmusen, E. 1994. *Games and Information: An Introduction to Game Theory*. Oxford: Blackwell. Fourth Edition, 2006.

Raux, A., Langner, B., Black, A. Eskenazi, M. 2005. Let's Go Public! Taking a Spoken Dialog System to the Real World. *Proceedings of Interspeech 2005*. On-line available at: http://www.cs.cmu.edu/~awb/papers/is2005/IS051938.PDF.

Reeves, N. Nass C. 1996. *The Media Equation: How People Treat Computers, Television, and New Media like Real People and Places*. New York: Cambridge University Press.

Reichman, R. 1985. *Getting Computers to Talk Like You and Me. Discourse Context, Focus, and Semantics (An ATN Model)*. Cambridge, Mass.: MIT Press.

Reiter, E., Dale, R. 2000, *Building Natural-Language Generation Systems*. Cambridge: Cambridge University Press.

Reithinger, N., Maier, E. 1995. Utilizing Statistical Dialogue Act Processing in Verbmobil. *Proceedings of the 33rd Annual Meeting of ACL*, pp. 116–21.

Rendell, L., Cho, H. 1990. Empirical Learning as a Function of Concept Character. *Machine Learning*, **5**: 267–298.

Rich, C. C., Sidner, C. L., Lesh, N., Garland, A., Booth, S., Chimani, M. 2005. DiamondHelp: A Collaborative Task Guidance Framework for Complex Devices. *Proceedings of the 20th AAAI Conference and the 17th Innovative Applications of Artificial Intelligence Conference*, pp. 1700–1. Pittsburgh: AAAI Press/The MIT Press.

Ries, K. 1999. HMM and Neural Network Based Speech Act Detection. In *Proceedings of IEEE international Conference on Acoustics, Speech, and Signal Processing (ICASSP)*, pp. 497–50. Also available on-line: citeseer.nj.nec.com/ries99hmm.html.

Rieser, V., Lemon, O. 2006. Using Machine Learning to Explore Human Multimodal Clarification Staregies. *Proceedings of the 21st International Conference on Computational Linguistics and 44th Annual Meeting of the Association for Computational Linguistics (COLING/ACL)*, Sydney, pp. 659–666.

Rizzolatti, G., Craighero, L. 2004. The Mirror-neuron System. *Annual Review of Neuroscience*, **27**: 169–192.

Roulet, E. 1986. Complétude Interactive et Mouvements Discursifs. *Cahiers de Linguistique Francaise*, **7**: 189–206.

Roy, N., Pineau, J., Thrun, S. 2000. Spoken Dialog Management for Robots. *Proceedings of the 38th Annual Meeting of the Association for Computational Linguistics (ACL-2000)*, pp. 93–100.

Rudnicky, A. I., E. Thayer, P. Constantinides, C. Tchou, R. Shern, K. Lenzo, W. Xu, A. Oh 1999. Creating natural dialogs in the Carnegie Mellon Communicator System. *Proceedings of the 6th European Conference on Speech Communication and Technology (Eurospeech-99)*, pp. 1531–1534. Budapest, Hungary.

Russel, S., Norvig, P. 2003. *Artificial Intelligence – A Modern Approach* (2nd Edition). Prentice Hall.

Sacks, H. 1992. *Lectures on Conversation, Volumes I and II*. Edited by G. Jefferson with Introduction by E.A. Schegloff, Oxford: Basil Blackwell.

Sacks, H., Schegloff, E. A., Jefferson, G. 1974. A simplest systematics for the organization of turn-taking for conversation. *Language* **50**: 696–735.

Sadek, D. 1999. Design Considerations on Dialogue Systems: from Theory to Technology – the case of Artimis. *Proceedings of the ESCA workshop on Interactive Dialogue Systems*, Kloster Irsee, Germany.

Sadek, D. 2005. ARTIMIS Rational Dialogue Agent Technology: An Overview. In *Multi-Agent Programming – Languages, Platforms and Applications*, Volume 15, pp. 217–243. US: Springer.

Sadek, D., Bretier, P., Panaget, F. 1997. ARTIMIS: Natural Dialogue Meets Rational Agency. *Proceedings of the Fifteenth International Joint Conference on Artificial Intelligence (IJCAI 97)*, Nagoya, Japan. Morgan Kaufmann, pp. 1030–1035.

Sahlgren, M. 2005. An Introduction to Random Indexing. *Proceedings of the Methods and Applications of Semantic Indexing Workshop*. The 7th International Conference on Terminology and Knowledge Engineering, TKE 2005, Copenhagen, Denmark.

Samuel, K., Carberry, S., Vijay-Shanker, K. 1998. Dialogue Act Tagging with Transformation-Based Learning. *Proceedings of the 36th Annual Meeting of the Association for Computational Linguistics and the 17th International Conference on Computational Linguistics (ACL-COLING)*, pp. 1150–6.

Scha, R., Polanyi, L. 1988. An Augmented Context-Free Grammar for Discourse. *Proceedings of the 12th International Conference on Computational Linguistics (COLING-88)*, pp. 22–7.

Schank, R. C., Abelson, R. P. 1977. *Scripts, Plans, Goals, and Understanding*. Hillsdale, New Jersey: Lawrence Erlbaum Associates.

Schatzmann, J., Stuttle, M. N., Weilhammer, K., Young, S. 2005: Effects of the user model on the simulation-based reinforcement-learning of dialogue strategies; IEEE ASRU workshop Automatic Speech Recognition and Understanding.

Scheffler, K. Young, S. 2000. Probabilistic Simulation of Human-Machine Dialogues. *Proceedings of the IEEE ICASSP*, pp. 1217–20, Istanbul, Turkey.

Scheffler, K. Young, S. 2002. Automatic Learning of Dialogue Strategy using Dialogue Simulation and Reinforcement Learning. *Proceedings of Human Language Technology*, pp. 12–18.

Schegloff, E. A. H. Sacks 1973. Opening up Closings. *Semiotica VIII*, **4**: 290–327.

Shneiderman, B. 1998. *Designing the User Interface: Strategies for Effective Human-Computer Interaction*. (3rd Edition). Reading, Mass.: Addison-Wesley Publishers.

Searle, J. R. 1969. *Speech acts: An essay in the philosophy of language*. Cambridge: Cambridge University Press.

Searle, J. R. 1979. *Expression and Meaning: Studies in the theory of Speech Acts*. Cambridge: Cambridge University Press.

Searle, J. 1980. Minds, Brains and Programs. *Behavioral and Brain Sciences* **3**(3): 417–57.

Seneff, S., Hurley, E., Lau, R., Pao, C., Schmid, P., V. Zue 1998. GALAXY-II: A reference architecture for conversational system development. *Proceedings of the 5th International Conference on Spoken Language Processing (ICSLP 98)*. Sydney, Australia.

Seneff S., R. Lau, J. Polifroni 1999. Organization, Communication, and Control in the Galaxy-II Conversational System. *Proceedings of the 6th European Conference on Speech Communication and Technology (Eurospeech-99)*, pp. 1271–4. Budapest, Hungary.

Sgall, P., Hajicova, E., Benesova, E. 1973. *Topic, Focus and Generative Semantics*. Kronberg Taunus: Scriptor Verlag.

Shriberg, E., Ferrer, L., Kajarekar, S., Venkataraman, A., Stolcke, A. 2005. Modeling prosodic feature sequences for speaker recognition. *Speech Communication*, **46**(3–4): 455–72. Special Issue on Quantitative Prosody Modelling for Natural Speech Description and Generation.

Sidner, C. L., Boettner, C, Rich, C. 2000. Lessons Learned in Building Spoken Language Collaborative Interface Agents. *In ANLP/NAACL 2000 Workshop on Conversational Systems - Volume 3, Seattle, Washington*, pp. 1–6.

Simpson, A., Fraser, N. F. 1993. Black box and Glass box Evaluation of the SUNDIAL System. *Proceedings of the 3rd European Conference on Speech Communication and Technology*, pp. 1423–6.

Sinclair, J. M., Coulthard, R. M. 1975. *Towards an Analysis of Discourse: The English used by teacher and pupils*. Oxford: Oxford University Press.

Singh, S., Kearns, M., Litman, D. J., Walker, M. 2000. Reinforcement Learning for Spoken Dialogue Systems. *Advances in Neural Information Processing Systems 12 (NIPS)*, Cambridge, Mass.: MIT Press.

Sjöberg, C., Backlund, A. 2000. *Technology Foresight: Visions of Future Developments in Information and Communication Systems*. Swedish National Board for Industrial and Technical Development (NUTEK). http://www.jrc.es/pages/iptsreport/vol49/english/MET1E496.htm.

Skantze, G. 2005. Exploring Human Error Recovery Strategies: Implications for Spoken Dialogue Systems. *Speech Communication*, **45**(3): 325–341.

Small, S., Strzalkowski, T., Liu, T., Ryan, S., Salkin, R., Shimizu, N., Kantor, P., Kelly, D., Rittman, R., Wacholder, N. 2004. HITIQA: Towards Analytical Question Answering. *Proceedings of the COLING 2004*, pp. 1291–7, Geneva, Switzerland.

Smith R. W. 1998. An Evaluation of Strategies for Selectively Verifying Utterance Meanings in Spoken Natural Language Dialog. *International Journal of Human-Computer Studies*, **48**: 627–47.

Smith R. W., Gordon, S. A. 1997. Effects of Variable Initiative on Linguistic Behavior in Human-computer Spoken Natural Language Dialogue. *Computational Linguistics*, **23**(1): 141–68.

Smith R. W., Hipp, D. R. 1994. *Spoken Natural Language Dialog Systems - A practical approach*. Oxford: Oxford University Press.

Sperber, D., Wilson, D. 1995. *Relevance: Communication and Cognition* (2nd Edition) Oxford: Blackwell.

Steedman, M. 1991. Structure and Intonation. *Language*, **67**(2): 260–96.

Steedman, M. 2000. Information Structure and the Syntax-Phonology Interface. *Linguistic Inquiry*, **31**(4): 649–89.

Steels, L. 2003. Evolving Grounded Communication for Robots. *Trends in Cognitive Science*, **7**(7): 308–12.

Steels, L. 2004. The Evolution of Communication Systems by Adaptive Agents. In Alonso, E., Kudenko, D., Kazakov, D. (Eds.), *Adaptive Agents and Multi-Agent Systems*, pp. 125–140. Springer-Verlag.

Stenström, A. B. 1994. *An Introduction to Spoken Interaction*. London: Longman.

Stent, A. Dowding, J., Gawron, J. M., Owen-Bratt, E., Moore, R. 1999. The CommandTalk Spoken Dialogue System. *Proceedings of the 37th Annual Meeting of the Association for Computational Linguistics*, pp. 20–6.

Stolcke, A., Ries, K., Coccaro, N., Shriberg, E., Bates, R., Jurafsky, D., Taylor, P., Martin, R., Van Ess-Dykema, C., Meteer, M. 2000. Dialogue Act Modeling for Automatic Tagging and Recognition of Conversational Speech. *Computational Linguistics*, **26**(3): 339–73.

Stork, D. (Ed.) 1998. *HAL's Legacy: 2001's Computer as Dream and Reality*. MIT Press.

Strube, M. 1998. Never Look Back: An Alternative to Centering. In *Proceedings of COLING-ACL'98: 36th Annual Meeting of the Association for Computational Linguistics and 17th International Conference on Computational Linguistics*, Montreal, pp. 1251–7.

Suchman, L., Blomberg, J., Orr, J., Trigg, R. 1999. Reconstructing Technologies as Social Practice. In Lyman, P. and Wakeford, N. (Eds.) Special issue of the American Behavioral Scientist on Analysing Virtual Societies: *New Directions in Methodology*, **43**(3): 392–408.

Suhm, B., Geutner, P., Kemp, T., Lavie, A., Mayfield, L., McNair, A. E., Rogina, I., Schultz, T., Sloboda, T., Ward, W., Woszczyna, M., Waibel, A. 1995. JANUS: Towards Multilingual Spoken Language Translation. *Proceedings of ARPA Workshop on Speech and Natural Language Technology (SLT-1995)*, Austin, Texas, pp. 185–189.

Swerts, M., Hirschberg, J., Litman, D. 2000. Correction in Spoken Dialogue Systems. *Proceedings of the International Conference on Spoken Language Processing (ICSLP-2000)*, Beijing, vol. 2, pp. 615–618.

Swerts, M., Geluykens R. 1994. Prosody as a Marker of Information Flow in Spoken Discourse. *Language and Speech*, **37**(1): 21–43.

Swerts, M., Krahmer, E. 2005. Audiovisual Prosody and Feeling of Knowing. *Journal of Memory and Language*, **53**: 81–94.

Takezawa, T., Morimoto, T., Sagisaka, Y., Campbell, N., Iida, H., Sugaya, F., Yokoo, A., Yamamoto, S 1998. A Japanese-to-English speech translation system: ATR-MATRIX. *Proceedings of the International Conference on Spoken Language Processing (ICSLP-1998)*, Sydney, Australia, pp. 957–960.

Tao, J. H., Tan, T. N. (Eds.) 2008. *Affective Information Processing, Science+Business Media LLC*, Springer-Verlag, London.

Terken J, Hirschberg J. 1994. Deaccentuation of Words Representing "Given" Information: Effects of Persistence of Grammatical Function and Surface Position. *Language and Speech*, **37**: 125–45.

Thagard, P. 2007. I Feel Your Pain: Mirror Neurons, Empathy and Moral Motivation. *Journal of Cognitive Science*, **8**: 109–136.

Theobalt, C., Bos, J., Chapman, T., Espinosa-Romero, A., Fraser, M., Hayes, G., Klein, E., Oka T., Reeve, R.. 2002. Talking to Godot: Dialogue with a mobile robot. *Proceedings of IEEE/RSJ International Conference on Intelligent Robots and Systems (IROS 2002)*, pp. 1338–43.

Thompson, E. 2001. Empathy and consciousness. *Journal of Consciousness Studies*, **8**: 1–32.

Tomasello, M. 1992. *First Verbs: A case study of early grammatical development*. Cambridge: Cambridge University Press.

Traum, D. R. 1994. A Computational Theory of Grounding in Natural Language Conversation. PhD Thesis, Department of Computer Science, University of Rochester, Rochester, NY. Technical Report TR-545.

Traum, D. 1999. Computational Models of Grounding in Collaborative Systems. In *Working Notes of AAAI Fall Symposium on Psychological Models of Communication*, pp. 124–31.

Traum, D. R. 2000. 20 Questions on Dialogue Act Taxonomies. *Journal of Semantics*, **17**: 7–30.

Traum, D. R., Allen, J. F. 1994. Discourse obligations in dialogue processing. *Proceedings of the 32nd Annual Meeting of the Association for Computational Linguistics*, pp. 1–8.

Traum, D. R., Rickel, R. 2002. Embodied Agents for Multi-party Dialogue in Immersive Virtual Worlds. *Proceedings of The First International Joint Conference on Autonomous Agents & Multiagent Systems (AAMAS)*, 15–19 July, *Bologna*, Italy, pp. 766–73.

Traum, D., Rickel, J., Gratch, J., Marsella, S. 2003 Negotiation over Tasks in Hybrid Human-Agent Teams for Simulation-Based Training. *Proceedings of the Second International Joint Conference on Autonomous Agents and Multi-Agent Systems*, Melbourne, Australia, pp. 441–8, July, 2003.

Turing, A. M. 1950. Computing Machinery and Intelligence. *Mind. A Quarterly Review of Psychology and Philosophy*. **59**(236): 433–60.

Turk, M., Robertson, G. (Eds.) 2000. Perceptual User Interfaces. *Communications of the ACM Special Issue*, **43**(3): 32–70.

Turunen, M. Hakulinen, J., Räihä, K-J., Salonen, E-P., Kainulainen, A., Prusi, P., 2005. An architecture and applications for speech accessibility systems.. *IBM Systems Journal,*, **44**(3): 485–504.

Tversky, B. 2000. Multiple Mental Spaces. *Plenary talk at the Int Conf on Rationality and Irrationality*, Austria. Available at: www-psych.stanford.edu/~bt/space/papers/rationality.pdf

Vallduví, E. 1995. Structural Properties of Information Packaging in Catalan. In Kiss, K. (Ed.) *Discourse Configurational Languages*. Oxford: Oxford University Press, pp. 122–52.

Vallduví, E., Engdahl, E. 1996. The Linguistic Realization of Information Packaging. *Linguistics* **34**: 459–519.

Vallduví, E., Vilkuna, M. 1998. On Rheme and Kontrast. In Culicover, P. and McNally, L. (Eds.) *The Limits of Syntax*. New York: Academic Press, pp. 79–109.

Van Kuppevelt, J. 1995. Discourse Structure, Topicality and Questioning, *Journal of Linguistics* **31**: 109–47.

VoiceXML Forum. Voice eXtensible Markup Language VoiceXML, Version 1.00. http://www.voicexml.org.

Von Wright, G. H. 1971. *Explanation and Understanding*. London: Routledge, Kegan and Paul.

Wahlster, W. (Ed.) 2000. *Verbmobil: Foundations of speech-to-speech translation*. Heidelberg, Berlin: Springer-Verlag.

Wahlster, W. (Ed.) 2004. *SmartKom – Foundations of Multimodal Dialogue Systems*. Heidelberg, Berlin: Springer-Verlag.

Wahlster, W. Marburger, H., Jameson, A., Busemann, S. 1983. Over-answering Yes-no Questions: Extended Responses in a NL Interface to a Vision System. *Proceedings of the 8th International Joint Conference on Artificial Intelligence (IJCAI'83)*, pp. 643–6, Karlsruhe.

Wahlster, W., Reithinger, N., Blocher, A. 2001. SmartKom: Multimodal Communication with a Life-Like Character. *Proceedings of the 7th European Conference on Speech Communication and Technology (Eurospeech 2001)*. Aalborg, Denmark.

Walker, M. A. 1989. Evaluating Discourse Processing Algorithms. *Proceedings of the 27th Annual Meeting of the Association of Computational Linguistics*. pp. 251–61.

Walker, M. 2000. An Application of Reinforcement Learning to Dialogue Strategy Selection in a Spoken Dialogue System for Email. *Journal of Artificial Intelligence Research*, **12**(12): 387–416.

Walker, M. A., Fromer, J. C., Narayanan, S. 1998. Learning Optimal Dialogue Strategies: A Case Study of a Spoken Dialogue Agent for Email. *Proceedings of the 36th Annual Meeting of the Association for Computational Linguistics and 17th International Conference. on Computational Linguistics*, pp. 1345–1352.

Walker, M. A., Hindle, D., Fromer, J., Di Fabbrizio, G., Mestel, G. 1997a. Evaluating Competing Agent Strategies for a Voice Email Agent. *Proceedings of the 5th European Conference on Speech Communication and Technology*.

Walker, M., Joshi, A. K., Prince, E. F. (Eds.) 1998. *Centering Theory in Discourse*. Oxford: Clarendon Press.

Walker, M. A., Langkilde, I., Wright, J., Gorin, A., Litman, D. J. 2000. Learning to Predict Problematic Situations in a Spoken Dialogue System: Experiments with How May I Help You? *Proceedings of the 1st Meeting of the North American Chapter of the Association for Computational Linguistics (NAACL-2000)*, Seattle, US, pp. 210–217.

Walker, M., Litman, D. J., Kamm, C. A., Abella, A. 1997b PARADISE: A Framework for Evaluating Spoken Dialogue Agents. *Proceedings of the 35th Annual Meeting of the Association for Computational Linguistics and 8th Conference of the European Chapter of the Association for Computational Linguistics (ACL-97/EACL-97)*, Madrid, Spain, pp. 271–280.

Walker, M., Litman, D. J., Kamm, C. A., Abella, A. 1998. Evaluating Spoken Dialogue Agents with PARADISE: Two Case Studies. *Computer Speech and Language*, **12**(3): 317–47.

Wallace, M. D., Anderson, T. J. 1993. Approaches to Interface Design. *Interacting with Computers*, **5**(3): 259–78.

Ward, N., Tsukahara, W. 2000. Prosodic Features which Cue Back-channel Responses in English and Japanese. *Journal of Pragmatics*, **23**: 1177–1207.

Wasinger, R., Oliver, D., Heckmann, D., Braun, B., Brandlarm, B., Stahl, C. 2003. Adapting Spoken and Visual Output for a Pedestrian Navigation System, based on given Situational Statements. Workshop on Adaptivity and User Modelling in Interactive Software Systems (ABIS), pp. 343–346.

Watkins, C. J. C. H, Dayan, P. 1992. Q-learning. *Machine Learning*, **8**(3): 279–92

Weinschenk, S. Barker, D. 2000. *Designing Effective Speech Interfaces*. London: John Wiley and Sons, Ltd.

Weiser, M. 1991. The Computer for the Twenty-First Century. *Scientific American*, **265**(3): 94–104.

Weizenbaum, J. 1966. ELIZA – A Computer Program for the Study of Natural Language Communication between Man and Machine. *Communications of the Association for Computing Machinery*, **9**: 36–45.

Wermter, S., Arevian, G., Panchev, C. 2000. Towards Hybrid Neural Learning Internet Agents. In Wermter, S. & Sun, R. (Eds.) *Hybrid Neural Systems*. Springer Verlag, Berlin, pp. 158–174.

Wickens, C. D., Holland, J. G. 2000. *Engineering Psychology and Human Performance*. Upper Saddle River, New Jersey: Prentice-Hall.

Wilcock, G. 2001. Pipelines, Templates and Transformations: XML for Natural Language Generation. *Proceedings of the 1st NLP and XML Workshop*, Tokyo, pp. 1–8.

Wilcock, G. Jokinen, K. 2001b. Design of a Generation Component for a Spoken Dialogue System. *Proceedings of the Natural Language Pacific Rim Symposium (NLPRS-2001)*, pp. 545–50, Tokyo.

Williams, J., Poupart, P., Young, S. 2008 Partially Observable Markov Decision Processes with Continuous Observations for Dialogue Management. In Dybkjaer, L. and Minker, W. (Eds.) *Recent Trends in Discourse and Dialogue*. Springer, pp. 191–217.

Williams, J. D., Young, S. 2006. Partially Observable Markov Decision Processes for Spoken Dialog Systems. *Computer Speech and Language* **21**(2): 393–422.

Winograd, T. 1972. *Understanding Natural Language*. New York: Academic Press.

Witten, I. H., Frank, E. (2005). *Data Mining: Practical machine learning tools and techniques* (2nd Edition). San Francisco: Morgan Kaufmann.

Wittgenstein, L. 1953. *Philosophical Investigations*. Oxford: Blackwell.

Wittgenstein, L. 1969. *On Certainty*. Edited by G. E. M. Anscombe and G. H. von Wright. Oxford: Blackwell.

Woods, W. A., Kaplan, R. N., Webber, B. N.. 1972. *The Lunar Sciences Natural Language Information System: Final Report*. BBN Report 2378, Cambridge, Mass.: Bolt Beranek and Newman Inc.

Yankelovich, N. 1996. How do Users Know What to Say? *ACM Interactions*, **3**(6): 32–43.

Young, S. 2000. Probabilistic Methods in Spoken Dialogue Systems.. In *Philosophical Transactions of the Royal Society (Series A)*, **58**(1769): 1389–1402.

Young, S. 2002. Talking to Machines (Statistically Speaking). Proceedings of the International Conference on Spoken Language Processing, (ICSLP) Denver, Colorado, pp. 9–16.

Young, S. L., Hauptmann, A. G., Ward, W. H., Smith, E. T., Werner P. 1989. High-level Knowledge Sources in Usable Speech Recognition Systems. *Communications of the ACM*, **32**(2): 183–94.

Zock, M., Sabah, G. (Eds.) 1988. *Advances in Natural Language Generation: An interdisciplinary perspective*. Pinter Publishers.

Zue, V. 1997. Conversational Interfaces: Advances and Challenges. *Proceedings of Eurospeech 97*, Rhodes, Greece, pp. KN 9–18.

Index

Constructive Dialogue Modelling   Kristiina Jokinen
© 2009 John Wiley & Sons, Ltd